Springer Proceedings in Business and Economics

More information about this series at http://www.springer.com/series/11960

Monika Bąk
Editor

Transport Development Challenges in the Twenty-First Century

Proceedings of the 2015 TranSopot Conference

Springer

Editor
Monika Bąk
Faculty of Economics, Chair of Comparative Research
of Transport Systems
University of Gdańsk
Gdańsk, Poland

ISSN 2198-7246 ISSN 2198-7254 (electronic)
Springer Proceedings in Business and Economics
ISBN 978-3-319-26846-0 ISBN 978-3-319-26848-4 (eBook)
DOI 10.1007/978-3-319-26848-4

Library of Congress Control Number: 2016934076

Printed on acid-free paper

This Springer imprint is published by Springer Nature
The registered company is Springer International Publishing AG Switzerland

Contents

Introduction ... 1
Monika Bąk and Michał Suchanek

International Cooperation in the EU Transport Research: Experiences and Barriers ... 3
Monika Bąk

Method for Assessing Rail Transport Competitiveness in Poland and the United Kingdom ... 17
Piotr Jurewicz and Daniel Kaszubowski

The Impact of Public Procurement System on Concentration of the Rail Freight Market in Poland ... 31
Grzegorz Krawczyk

Local Government, The Hostage or Playmaker in Competition Between Modes of Transport: Examples from Local Freight Transport in Lower Silesia ... 39
Krzysztof Lewandowski

EU Transport Policy Failure: The Case of Germany's Mindestlohngesetz ... 51
Wojciech Paprocki

Costs and Benefits of High Speed Rail Integration in Europe ... 67
Anna Urbanek

Objectives and Strategies of Sustainable Urban Mobility Planning in the City of Krakow ... 77
Maciej Michnej and Tomasz Zwoliński

The Significance of Pedestrian Mobility as Shown by the Example of the City of Gdynia ... 85
Marcin Wołek

Cars and Urban Travel 93
Katarzyna Hebel

Comparison of Road Safety Policy Objectives in Poland and in the European Union 103
Monika Paradowska

Public–Private Partnership and the Development of Transport Infrastructure in Poland: The Analysis of Critical Success Factors 125
Beata Zagozdzon

Transportation Services as Specific Logistics Projects for Oversized Cargo in Poland 139
Iwona Pisz and Iwona Łapuńka

The Impact of Traffic Parameter Assessment on Noise Emission 161
Justyna Sordyl

Identification of Behavioral Changes in Transport as a Means for Decarbonisation of the Economy 171
Urszula Motowidlak

Pro-Innovative Systems of Decision-Making in TSL Enterprises: Behavioural Conditions 181
Joanna Fryca-Knop

Simulation of Technical and Economical Processes as an Initial Phase of Electric Bus Fleet Implementation to Operation in Urban Public Transport Company 193
Krzysztof Krawiec

The Impact of Organizational Culture on Bicycle Commuting Frequency: The Research Based on Example of Three IT companies 201
Romanika Okraszewska

Risk Factor Classification GEMIO in the Planning Phase of Logistic Project Management 211
Dorota Książkiewicz

Use of Economic and Econometric Analysis in the Financial Standing Diagnosis of Haulage Enterprises, Case Study: Trans Polonia S.A. 221
Andrzej Letkiewicz and Michał Suchanek

Koleje Śląskie as an Example of H. Mintzberg's Structural Configuration Theory Application in Designing Organizational Structure of a Railway Operator 231
Robert Tomanek

Introduction

Monika Bąk and Michał Suchanek

Transport is undoubtedly an extremely important field of human activity. Its development is a vital part of every political and economic strategy both on a local and on an international level. Achieving transport efficiency enables various economies to unlock their growth potential, create jobs and bring wealth to communities. Thus, it is not only a promising business field but also a crucial matter of scientific research. With that point in mind, 'TranSopot 2015 Conference: Transport Development Challenges in the Twenty-First Century' has been held from 25 to 27 May 2015 at the University of Gdańsk, Faculty of Economics. The proceedings of the conference are presented in this book.

The book is divided into four parts, each focusing on different transport aspect.

The first part contains articles in which international, especially European, aspects of transport policy are considered. Some of the articles present a critical view of certain, unsuccessful cases of transport policies whose effects turned out to be quite different from their assumptions.

The second part focuses on the transport activity in light of the sustainable development idea. The articles present problems connected with road safety in Poland and in the European Union but also challenges in sustainable transport planning, especially in metropolitan areas.

The third part of the book includes articles analysing different innovations in transport development. The authors present innovations in railway transport,

M. Bąk (✉)
Chair of Comparative Research of Transport Systems, Faculty of Economics, University of Gdańsk, Jana Bażyńskiego 1A, 80-952 Gdańsk, Poland
e-mail: monika.bak@ug.gda.pl

M. Suchanek
Chair of Economics and Management of Transportation Companies, Faculty of Economics, University of Gdansk, Jana Bażyńskiego 1A, 80-952 Gdańsk, Poland
e-mail: m.suchanek@ug.edu.pl

M. Bąk (ed.), *Transport Development Challenges in the Twenty-First Century*, Springer Proceedings in Business and Economics,
DOI 10.1007/978-3-319-26848-4_1

logistics planning and noise emission and also identify different behavioural changes vital for the decarbonisation of the economy.

The fourth part of the book introduces various aspects of functioning of organisations which perform transport activity. It is worth mentioning that the articles present both public and private transport companies. Different facets of management are presented: from logistic and risk planning, simulation through work organisation and organisational culture to economic and econometric analysis which can be used to diagnose transport enterprises.

Due to the fact that the articles cover a wide range of transport topics, not only from the theoretical standpoint but also from the practical one, these proceedings can be interesting for a diversified audience. Scientists can find new and interesting aspects of transport research, for transport practitioners methods useful in planning and managing their enterprises can provide added value, while policymakers can take advantage of cases of both successful and appalling transport policies. These proceedings can be seen as a review of the transport development challenges in the twenty-first century.

International Cooperation in the EU Transport Research: Experiences and Barriers

Monika Bąk

Abstract Transport research is crucial for the development of transport policy and solving current and future problems of transport market functioning and its impact on social and economic life as well as on the environment. Transport is global and at present it seems to be more and more important to develop a wider approach in transport research and to support international cooperation in this field. In this paper the relation between transport policy objectives and the EU strategy for international cooperation in the research and innovation (R & I) field has been analyzed. Additionally barriers of international cooperation in transport research are identified at the level of legal, institutional, financial, technological, and social and behavioral aspects. Some solutions to overcome these obstacles are also proposed.

Keywords International cooperation • Transport research • Stakeholders

Introduction

Transport sector meets numerous challenges such as minimalizing environmental threats, reducing risk of accidents, improving quality of service for passenger and goods, etc. As a global sector, transport is also diversified between countries in the context of technological development, legal background, and institutional framework. Innovations are needed in order to reduce barriers for further transport development. They can be developed and implemented in a specific geographical context but also at the international level. The second approach is especially important because it provides wider applications and ensures effectiveness to a significant degree. The objective of this article is to present the relation between transport policy objectives and the strategy of international cooperation in transport research at the EU level, to identify challenges and barriers of the research, and to provide some recommendations.

M. Bąk (✉)
Faculty of Economics, Chair of Comparative Research of Transport Systems, University of Gdańsk, Jana Bażyńskiego 1A, 80-952 Gdańsk, Poland
e-mail: monika.bak@ug.gda.pl

M. Bąk (ed.), *Transport Development Challenges in the Twenty-First Century*, Springer Proceedings in Business and Economics,
DOI 10.1007/978-3-319-26848-4_2

Objectives of the EU Policy of International Cooperation in Research vs. Transport Policy Priorities

As it is stated in the EC communication on enhancing and focusing EU international cooperation of September 2012 [15], international cooperation in research and innovation contributes to the broader policies of the Union, as reflected in the Europe 2020 strategy, in supporting the following objectives:

- Strengthening the Union's excellence and attractiveness in research and innovation and its economic and industrial competitiveness
- Tackling global societal challenges, such as food and energy security and climate change
- Supporting the Union's external policies

The question is if these objectives are in line with the priorities of the EU transport policy and can be helpful in realizing this policy. According to the White Paper on Transport of 2011 [19], a vision for a competitive and sustainable transport system requires achieving several goals:

- Growing transport and supporting mobility while reaching the 60 % emission reduction target
- An efficient core network for multimodal intercity travel and transport
- A global level-playing field for long-distance travel and intercontinental freight
- Clean urban transport and commuting

It is also mentioned that "implementing the above vision requires an efficient framework for transport users and operators, an early deployment of new technologies and the development of adequate infrastructure" which can be achieved through realizing strategic objectives of creating a single European transport area, innovating for the future—technology and behavior—and ensuring modern infrastructure and smart funding. One of the horizontal strategic instruments is also focusing on external dimension of the EU transport policy. Since transport is fundamentally international, most actions in the road map are linked to challenges related to the development of transport beyond the EU borders. External dimension as a horizontal issue is inevitable and includes all transport modes and almost all spheres of market functioning. It can be concluded then that the realization of transport objectives is closely linked to general EU international cooperation strategy.

Moreover the need to incorporate EU transport priorities into the subthemes and EU research policy should be stressed. As it is presented in Fig. 1, strengthening the Union's excellence and attractiveness in research and innovation as well as its economic and industrial competitiveness would be achieved through creating win–win situations and cooperating on the basis of mutual benefit. Instruments consist of enhancing R & D cooperation with the EU-associated countries and other countries worldwide but also of improving integration which helps to create wider market. In transport sector the impact of this policy can be perceived in all aspects of priorities' realization. A single European transport area could be further developed with a

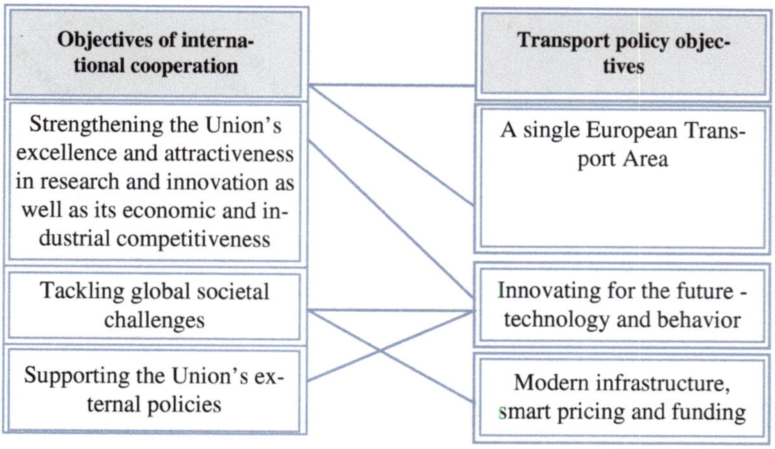

Fig. 1 Relation between EU high-level objectives of international cooperation and transport policy objectives [own elaboration]

closer integration of the research within and also outside the EU. New possibilities will emerge in the field of innovating for the future transport systems in both technology and behavior and creating modern infrastructure, smart pricing, and funding. If one assumes further enlargements or looks at the international dimension of transport modes, it appears to be clear that it is impossible to complete the Single European Sky, the European maritime transport space, or liberalized road freight market without cooperation with third countries (including research cooperation).

Tackling global societal challenges requires developing and deploying effective and innovative solutions in both technology and behavior. In general innovating for the future is very closely connected with research policy and international cooperation aiming at transferability of best practices or new concepts' elaboration. Additionally, new and modern solutions in transport infrastructure can be developed through international cooperation as well as smart pricing, especially for infrastructure, and new funding schemes are global challenge for the research. The achievements of modern infrastructure and smart funding goals can be accelerated by the cooperation in research regarding practical solutions, their implementation potential and barriers, etc.

Supporting the Union's external policies is in line with the formation of innovations in transport. It could be achieved through enhancing R & D cooperation with two broad stakeholder groups—EU-associated and neighboring countries and other countries worldwide.

Advantages of international cooperation in research can be perceived from the perspective of international science, technology, and innovation agreements which are strongly supported by the European Commission [see, e.g. 2, 5, 13]. According to the European Commission study, potential advantages can be grouped into a narrow and a broad science and technology and innovation agreement paradigm [5]. The first one is related only to quality, scope, and critical mass in STI (science, technology, and innovation) by linking resources and knowledge with resources and knowledge in other coun-

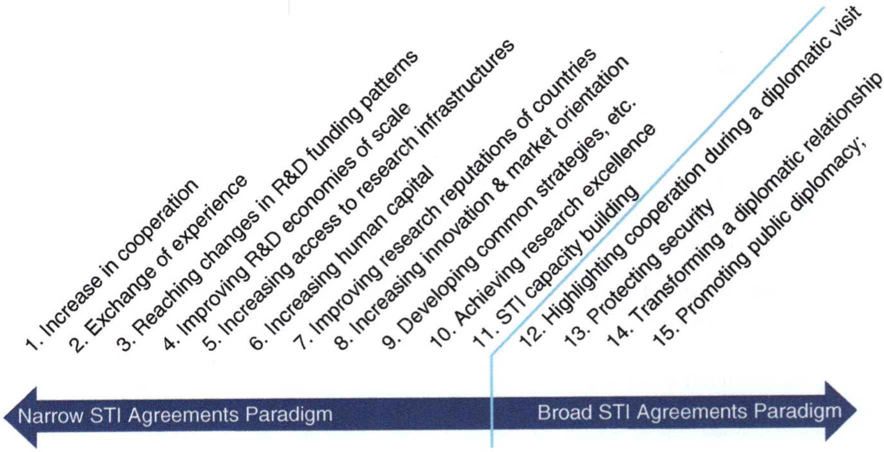

Fig. 2 Reasons for signing science and technology agreements grouped in two paradigms [5]

tries. The second paradigm also includes nonscience policy objectives. In the broad paradigm, signing a bilateral STI agreement becomes a means to reach policy ends outside the realm of science, technology, and innovation. The figure below presents both paradigms and the respective reasons for signing a bilateral STI agreement (Fig. 2).

From the perspective of transport sector, this impact is also significant and can be perceived from both the policymakers' and researchers' point of view. Stronger multilateral relations in research would improve mutual trust and policy networks and partnership in general which is especially important in such an international sector as transport. Additionally international cooperation in transport research improves implementation and enforcement processes and ensures better insides into best and worst practices in domestic transport systems. Researchers can expect such benefits as creating new knowledge and focusing on innovations through better quality research team. This results in innovative solutions in transport technology and transport system development. International cooperation provides better access to expertise, skills, and research areas as well as efficient dissemination of research results and best practice examples. That seems to be an important factor in the area of smart pricing and funding in transport. Within international cooperation complex, scientific and technical problems, for example, regarding trans-European networks or a single European transport area, can be tackled in a faster and better way. Additionally, there is an increased access to funding for research.

The European Union Initiatives in Promoting and Supporting International Cooperation in Transport Research

At the EU level, some initiatives aiming at improving international cooperation in research are developed and supported. The major instrument of the EU research policy is Framework Programmes for Research and Technological Development, realized from 1984 (the First Framework Programme) until now—Eighth Framework Programme named Horizon 2020.

The origins of the Framework Programme are in European efforts to close a perceived "technology gap," first with the USA and later with Japan, and to promote European competitiveness, especially in energy and information technology [2]. The changing nature of global competition and the progress of the European project toward closer union mean that the role of the Framework Programme has evolved. Initially, it was an effort to support European industrial competitiveness in a limited number of sectors by networking together and strengthening European technology development effort. It has evolved to become a larger and more powerful instrument for funding and coordinating scientific research as well as more industry-orientated technology efforts across Europe.

The selection of results from EU-funded projects that have shown great strides in innovation, including transport-oriented project, is presented in the brochure published by the European Commission [7]. But still it should be remembered that the changing global landscape (e.g., new players emerged in science, technology, and innovation researches like China) both creates opportunities and increases the need for strengthening internationalization of the research [11].

In 2000, since the launch of the European Research Area (ERA), Framework Programmes have become a component in a wider EU research and innovation policy. European Research Area ([21], see also [9]) is the instrument—area open to the world based on the internal market, in which researchers, scientific knowledge, and technology circulate freely. The objective of ERA is to strengthen the EU scientific and technological bases, including optimal transnational competition and cooperation through jointly addressing grand challenges. Additionally, the ERA external link consists of research activities, programmes, and policies in Europe which involve a transnational perspective, including cross-border cooperation.

At the EU level, some initiatives for dissemination information on international research are also developed. CORDIS should be mentioned here as the European Commission's primary public repository and portal to disseminate information on all EU-funded research projects and their results in the broadest sense (http://cordis.europa.eu/home_en.html [20]). Alternative information portal is at the website of the Directorate-General for Research and Innovation. Its mission is to develop and implement the European research and innovation policy with a view to achieving the goals of Europe 2020 and the Innovation Union [24].

As it was mentioned earlier, the most important instruments of the EU research policy are Framework Programmes. The specific objectives of the programs vary between funding periods. In FP6 and FP7, focus was still on technological research; in Horizon 2020 the focus is on innovation, delivering economic growth faster, and delivering solutions to end users that are often governmental agencies. Horizon 2020 is open to the participation of researchers from across the world. As more research and innovation is performed in international partner countries, it is crucial that Europe is able to access the best researchers and research centers worldwide. Not only does this provide sources of new ideas and expertise, but it is also important to ensure that European researchers are able to collaborate worldwide with the best in the field.

In 2014 the communication from the European Commission titled "Research and innovation as sources of renewed growth" [16] was published ("research and innovation"—R & I term is now commonly used at the EU level instead of former "research and development", R & D). In this communication additional investment is confirmed from the Union budget for R & I, but it is also clearly said that these funds must complement (not substitute for) investments by member states from both public and private sources. Though there is no special attention put on the international cooperation, it is clear that objectives stated here cannot be achieved without such initiatives.

Focusing on transport sector, research programs should be also mentioned. The previous one—Seventh Framework Programme—included the specific program "cooperation" and its thematic area "transport" (including aeronautics) with the objectives of the development of integrated, safer, "greener," and "smarter" pan-European transport systems for the benefit of all citizens and climate policy, respecting the environment and natural resources, and securing and further developing the competitiveness attained by the European industries in the global market [10].

It should be added that Framework Programmes are also targeted to third countries. In the seventh FP, Russia, China, and Japan and to a lesser extent Latin America, India, and industrialized countries were included.

In Horizon 2020 the so-called Transport Challenge is allocated a budget of 6339 million EUR [6] for the period 2014–2020 and will contribute to four key objectives, each supported by specific activities [25]:

- A resource-efficient transport that respects the environment by making aircraft, vehicles, and vessels cleaner and quieter to minimize transport system's impact on the climate and environment; by developing smart equipment, infrastructures, and services; and by improving transport and mobility in urban areas.
- A better mobility, less congestion, and more safety and security with a substantial reduction of traffic congestion and with a substantial improvement in the mobility of people and freight, by developing new concepts of freight transport and logistics and by reducing accident rates, fatalities, and casualties and improving security.
- A global leadership for the European transport industry by reinforcing the competitiveness and performance of European transport manufacturing industries and related services including logistic processes and retaining areas of European leadership (e.g. aeronautics).
- A socioeconomic and behavioral research and forward-looking activities for policy making. The aim is to support improved policy making which is necessary to promote innovation and meet the challenges raised by transport and the societal needs related to it.

In 2012 communication from the European Commission [14] on developing a European transport-technology strategy, it is confirmed that global environmental challenges need a coordinated global response, so clearly a need for international cooperation is formulated. Also, the commission postulates that transport research and innovation should be more focused. Therefore in the Work Programme 2014–2015 in the area of "Smart, green and integrated transport", the following activities are addressed by three calls for proposals [17]:

- Mobility for growth
- Green vehicles
- Small business and fast track innovation for transport

As indicated in the Specific Programme, the "activities will be organized in such a way as to allow for an integrated and mode-specific approach as appropriate." Therefore, the contents of the "Mobility for Growth" call have been organized as follows:

- Areas addressing mode-specific challenges (technical and socioeconomic): aviation, rail, road, and waterborne
- Areas addressing transport integration-specific challenges (technical and socioeconomic): urban, logistics, intelligent transport systems, and infrastructure
- Areas addressing cross-cutting issues: socioeconomic and behavioral research and forward-looking activities for policymaking

In transport sector there exist also some initiatives of the EU promoting international cooperation in research. Transport Research and Innovation Portal (TRIP), formerly known as the Transport Research Knowledge Centre (TRKC), should be mentioned. Here in-depth information about projects, thematic reports, and policy brochures are available [26].

EC research and innovation portal is also available for transport [23]. Policy directions for research are published there, and evaluations of the results of the research projects are available in the unified form of the EC publications:

- Communicating EU transport research Directorate-General for Research and Innovation [4]
- More efficient—less polluting. How 20 years of EU research cleaned up the internal combustion engine and made it drive better [8]
- Aeronautics and Air Transport Research. Project Synopses [1]
- Towards European Integrated Ocean Observation. Expert Group on Marine Research Infrastructures [12]

Finally, also international cooperation with other countries in the area of transport should be mentioned [22]. Successful specific cooperation actions were developed with Canada, China, Latin America, India, Japan, Russia, South Africa, Ukraine, and the USA in the period 2007–2011. Coordinated calls included external link with China and Russia in 2010.

Barriers of the International Transport Research

Advantages of international cooperation in the research will be reduced if existing and emerging barriers are not taken into consideration in planning strategies and implementation processes. Table 1 presents the proposal for classification of barriers of international cooperation in the research, their importance in transport research, as well as possible solutions to overcome them. Obstacles can be grouped in five areas representing legal, institutional, financial, technological, and social conditions.

Table 1 Classification of barriers of international cooperation in transport research

Type of barrier	Impact on the cooperation in research[a]		Importance for cooperation in transport research[b]	Possible solutions to overcome barriers
	Within the EU	With non-EU partners		
Legal/lack or inadequate legal regulations	–	v	++	Removing remaining legal constraint cooperation New concepts of cooperation Harmonization of rules Enforcement issues Statistical evidence and reporting
Institutional/ not well-adjusted institutional framework	v	v	+++	Involvement of different stakeholders, e.g., SME Cooperation in research between private and public sector Institutional framework Minimizing risks of conflicts and inefficiencies Ensuring coordination between different levels
Financial/ lack of or insufficient resources	v	v	+++	Additional financing sources Alternative financing of research Cooperation with industry Efficient pricing and financing New models of financing Dissemination of best practices and transferability of solutions
Technological/ insufficient technological support	–	v	+	New platforms for cooperation Implementation issues Best practices dissemination and transferability
Social and behavioral/lack of support from public opinion	v	v	+	Publicity for R & D cooperation Information campaigns Raising awareness and social acceptance for new solutions Ensuring nondiscrimination of some user groups Involving wide groups of stakeholders

[a]v barriers influence negatively cooperation on this level; – barriers do not influence negatively cooperation on this level

[b]+++ barriers are very important for cooperation in transport research; ++ barriers are important for cooperation in transport research; + barriers are not very important for cooperation in transport research

Legal barriers still play important role in the range of international cooperation between partners from differentiated countries, especially from the EU states and developing countries. Enforcement issues also differ between some countries which can be a constraint for the research. Removing existing legal obstacles and harmonization of rules are the solutions to be implemented in the cooperation in research, including transport field. Institutional framework is important mainly in order to involve private sector into research and minimizing risks of inefficiencies. Insufficient financial resources influence negatively the progress in the international cooperation in transport research. Developing new and alternative models of financing as well as dissemination of best practices in financing and transferability of these solutions can improve the situation. In the case of technology in transport area, barriers do not strongly influence international cooperation. But some improvements would be useful, for example, creation of new platforms for cooperation, especially with non-EU partners. Public support is needed for efficient international cooperation in research. Then raising awareness, involving wide groups of stakeholders and publicity, can help in achieving social acceptance.

In the report published by the EC including results of the study on "European Added Value of EU Science, Technology and Innovation actions and EU-Member State Partnership in international cooperation," three major barriers of international cooperation are identified both by EU and third-country researchers and stakeholders [13]:

- Lack of financial means to support co-funding mutual research undertakings
- Administrative burdens of organizing cooperation
- Lack of knowledge and information about other country's strengths and complementarities

It is also interesting that political barriers or geographical distance is not treated as a problem anymore. Still some strong administrative barriers of the international cooperation in research between the EU and third countries exist [3]. These barriers include the so-called EC bureaucratic barriers which involve obstacles related to constant changes of the rules and procedures, duration of project evaluation, payment delays, etc. The next group of barriers consists of institutional barriers at national level (e.g., lack of the country's lobbying skills at the level of EU administration, low scientific image of a country, etc.).

Barriers of international cooperation in research are not easy to overcome. But many efforts have been already done at the European Union and other country levels (see [13]). The most important aspect is elaborating and implementing research finding schemes at the EU level. They allowed to reduce barriers of cooperation through supporting networking, facilitating organization of the research projects, and fostering harmonization at legal and institutional levels.

Conclusions: Prioritization of the Research

When analyzing priorities and regarding the emphasis put on international cooperation objectives of the research policy of the European Union on the one hand and transport research strategies on the other, one should expect a very intensive progress in the realization of international projects and improving their positive effects. This is the case only to some extent. The EU instruments have been developed intensively, but many barriers still hinder the implementation of international research cooperation. Then it is a need of comprehensive and continuous analysis of future challenges for research and policy which should result from the state of the art and a review of existing gaps. Then the research needs can be translated into researchable questions. The final step should be a research prioritization, but this can be done only when criteria for prioritization are set.

Research on international cooperation can be prioritized mainly when taking into consideration its relevance to EU transport objectives and EU research policy. Moreover, a challenging task is to include additional research prioritization criteria, especially feasibility of research and potential impact. Additionally, the following criteria should also be taken into account:

- Potential for multiplicative effects
- Desirability of new research
- Additional to the EU funding opportunities

Prioritization of future research can be additionally considered more technically on two principal levels: strategic and operational. On strategic level the main challenges refer to:

- Necessity to apply specific approaches depending on the maturity of the cooperation, the country's scientific capacity
- Seeking for additional funding components (industry, SME, partner country, etc.)
- Guaranteeing visibility for the public and decision makers, especially facing concerns related to practicality, timing, cost, and impact of jointly produced innovative transport technologies, solutions, and legislative proposals

At operational level, concerning project and program management, the topics relate to:

- Timeliness and quality of information (solution: keep working on dissemination of practical information about EU's large-scale funding opportunities)
- Too many objectives (solution: many activities, few objectives, clear focus)
- Lack of explanation of obstacles and problematic areas (solution: project requirements to stress problems, not only justify the funding)
- Lack of directions for future research (solution: project requirements to provide advice on future work)

In addition to the above-discussed general objectives, one can also include the set of criteria mentioned in the Regulation of the European Parliament and of the Council, establishing Horizon 2020 ([18], see also [9]). As it is stated there, areas for engaging with third countries should be identified in a systematic and coherent manner on the basis of an analysis of the Union vis-à-vis the rest of the world in line with the following set of criteria:

- Research and innovation capacity, including investment, output, human resources, and infrastructure—that means research which would obviously benefit financially from being carried out jointly, after taking account of the additional costs of international collaboration. Research which contributes to the cohesion of the common market and which promotes the unification of European science and technology
- Risks of and opportunities for access to existing, new, or emerging markets and their impact on the EU competitiveness
- Contribution to the Union's international commitments, as reflected in the Millennium Development Goals, the post-2015 development framework, Rio + 20, G20 and the international objectives of sectoral policies—that means research which would achieve significant results in the whole of the community for problems on the international scale, owing to the complementary nature of national efforts
- The legal and administrative frameworks in place, among the international partners—that means research which leads when necessary also to the establishment of uniform laws and standards and where appropriate the member states, to engage in cooperation, also including lessons learned from previous cooperation

Transport sector is global and a need of international cooperation in transport research is obvious. Instruments supporting funding for mutual research have to be further developed to make them increasingly smart and efficient. Existing barriers of the cooperation, especially when they include third countries of the EU, should be reduced. Research on international cooperation should contribute to the wider objectives on transport integration based on three pillars: economic efficiency, social fairness, and environmental sustainability. From the perspective of European countries, the problems resulting from discrepancies between EU and non-EU countries are still unresolved. Then one recognizes the need to enhance R & D cooperation which finally should result in improving integration of the EU but also integration of transport sector in a wider meaning, as well as in the environmental protection, higher quality of transport services, and safety of transport sector.

References

Books, Series and Reports

1. Aeronautics and Air Transport Research. 7th Framework Programme 2007–2013. Project Synopses – Volume 2 Calls 2010 & 2011. Brussels: Directorate-General for Research and Innovation, European Commission (2012)
2. Arnold E., Mahieu B., Stroyan J., et al. Understanding the Long Term Impact of the Framework Programme Final Report to the European Commission. DG Research Under Framework Contract No. DG BUDG06/PO/01/LOT no. ABAC 101908. 2011. 98 p.

3. Barriers in research cooperation of WBC countries. Deliverable 3.16. WBC-INCO.NET (Coordination of Research Policies with the Western Balkan Countries. Project number: PL 212029 (2009)
4. Communicating EU transport research. Directorate-General Directorate-General for Research and Innovation – Transport. European Commission (2012)
5. Fikkers D.J, Horvat M.: Basic Principles for effective International Science, Technology and Innovation Agreements. Brussels: European Commission, Directorate-General for Research and Innovation, International Cooperation (2014)
6. HORIZON 2020 in brief. The EU Framework Programme for Research & Innovation European Commission, Directorate-General for Research and Innovation (2014)
7. Investing in European success. HORIZON 2020 Research and Innovation to boost growth and jobs in Europe. Brussels: Directorate-General for Research and Innovation, European Commission (2013)
8. More efficient – less polluting. How 20 years of EU research cleaned up the internal combustion engine, and made it drive better. Brussels: Directorate-General for Research & Innovation, European Commission (2012)
9. Recommendations on the Implementation of the ERA Communication by Member States and by the European Commission. Report of the Expert Group 2013. Brussels: Directorate-General for Research and Innovation, European Commission (2013)
10. Remotti L., Damvakeraki T., Ploder M., Sterner C. International Science and Technology Cooperation in the EU's 7th Framework Programme: the specific programme 'Cooperation' and its thematic areas. European Commission, Directorate-General for Research and Innovation. International Cooperation (2014)
11. Schwaag Serger S., Remoe S. International Cooperation in Science, Technology and Innovation: Strategies for a Changing World. Report of the Expert Group established to support the further development of an EU international STI cooperation strategy. Brussels: Directorate-General for Research and Innovation International Cooperation, European Commission (2012)
12. Towards European Integrated Ocean Observation. Expert Group on Marine Research Infrastructures. Final Report. Brussels: Directorate-General for Research and Innovation. EU Marine and Maritime Research Strategy. European Commission (2013)
13. Vullings W., Arnold E., Boekholt P., Horvat M., Mostert B., Rijnders-Nagle M. European Added Value of EU Science, Technology and Innovation actions and EU-Member State Partnership in international cooperation. Brussels: European Commission Directorate-General for Research and Innovation (2014)

Legislation

14. Communication from the Commission to the Council and the European Parliament: Research and innovation for Europe's future mobility. Developing a European transport-technology strategy. COM(2012) 501 final. Brussels (2012)
15. Communication from the Commission to the European Parliament, the Council, the European Economic and Social Committee and the Committee of the Regions: Enhancing and focusing EU international cooperation in research and innovation: A strategic approach. European Commission. COM (2012) 497 final. Brussels (2012)
16. Communication from the Commission to the European Parliament, the Council, the European Economic and Social Committee and the Committee of the Regions: Research and innovation as sources of renewed growth. COM(2014) 339 final. Brussels (2014)
17. HORIZON 2020 Work Programme 2014–2015 Smart, green and integrated transport. European Commission Decision C (2013) 8631 of 10 December 2013. Brussels (2013)

18. Regulation (EU) no 1291/2013 (EU, 2013) of the European Parliament and of the Council of 11 December 2013 establishing Horizon 2020 – the Framework Programme for Research and Innovation (2014–2020) and repealing Decision No 1982/2006/EC. Official Journal of the European Union L 347/104. Brussels (2013)
19. Roadmap to a Single European Transport Area – Towards a competitive and resource efficient transport system. White Paper. European Commission. 28.3.2011, COM (2011) 144 final. Brussels (2011)

Electronic Materials

20. CORDIS, http://cordis.europa.eu/home_en.html
21. European Research Area, http://ec.europa.eu/research/era/index_en.htm
22. International Transport Cooperation. Digital Agenda for Europe. A Europe 2020 initiatives, https://ec.europa.eu/digital-agenda/node/262
23. Research & Innovation. Transport. European Commission, http://ec.europa.eu/research/transport/index_en.htm
24. Research and Innovation portal. European Commission, http://ec.europa.eu/research/index.cfm
25. Smart, Green and Integrated Transport, http://ec.europa.eu/programmes/horizon2020/en/h2020-section/smart-green-and-integrated-transport
26. Transport Research & Innovation Portal (TRIP), http://www.transport-research.info/web/

Method for Assessing Rail Transport Competitiveness in Poland and the United Kingdom

Piotr Jurewicz and Daniel Kaszubowski

Abstract Both in Poland and the United Kingdom, as well as worldwide, railways have to compete with other modes of transport, and the rail industry seeks possibilities of achieving a competitive advantage over them. Thus, there is a need for methods for assessment of rail transport competitiveness and detection of areas for improvement. This paper presents a method developed for the Polish regional transport market and verified according to the requirements of the British market. A numerical competitiveness evaluation, called Rail Transport Competitiveness Index, considers four criteria: journey duration, ticket price, service frequency and number of changes. Surveys in Poland and the United Kingdom have been carried out in order to assign relevant weights of importance to each criterion. The method has been applied on 18 routes in Pomerania Province, to create a map of rail transport competitiveness in the region. After necessary modifications, considering dynamic fares, it has been used for Milton Keynes–London route. Its universal nature makes it useful also for other countries, possibly after minor modifications.

Keywords Railway competitiveness • Transport competitiveness • Public transport • Competition • Market • Competitive advantage • Rail assessment

Introduction

In the reality of competition on the transport market, the rail industry seeks for possibilities of achieving a competitive advantage over other modes of transport. According to Smoliński [1], such approach to the issue of competitiveness is crucial, as the open market requires transport companies to be able to build and

P. Jurewicz (✉)
Research and Development Department, Rail Safety and Standards Board, London, UK
e-mail: pkjurewicz@gmail.com

D. Kaszubowski
Highway Engineering Department, Gdańsk University of Technology, Gdańsk, Poland
e-mail: daniel.kaszubowski@wilis.pg.gda.pl

M. Bąk (ed.), *Transport Development Challenges in the Twenty-First Century*,
Springer Proceedings in Business and Economics,
DOI 10.1007/978-3-319-26848-4_3

constantly keep the competitive advantage over others and to constantly improve the quality of service. Thus, there is a need for methods allowing to assess competitiveness of rail transport and detect the areas for improvement, which may be useful for planning of changes and development projects. Situation in both analysed countries regarding the role of rail passenger transport differs significantly. In Poland rail transport faces significant decrease when passenger transport modal split is considered in passenger-kilometre travelled. Rail transport share decreased from 8 % in 2004 to 4.8 % in 2014, being as high as 15.5 % in 1995. In the same period in the United Kingdom, rail increased its share from 5.7 to 8.2 %, while in 1995 it was only 4.4 %. This indicates that rail in both countries faces different challenges. In Poland it has to seek for improvement of competitiveness against fast growing road transport to stop the market share deterioration. In the United Kingdom, market opportunities should be identified to foster a process of modal split change. Moreover, recent research by Rosik et al. [2] regarding road transport accessibility in Poland revealed that 10 years of EU-funded road investments improved the cohesion on national network, but neglected regional accessibility. In this situation rail should take an effort to challenge road transport at the regional level to fill this gap providing better offer for passengers to gain long-term competitive advantage.

A universal method for assessing competitiveness of rail transport in comparison to other modes of public transport has been developed by the authors as a part of M.Sc. thesis [3]. The method considers various criteria of choosing a mode of transport, together with relevant weights of importance assigned to each criterion. It has been applied to evaluate rail transport competitiveness in Pomerania Province, Poland, and used to create a map of rail transport competitiveness in the region. In this paper we present the method used for the Polish regional transport market and the results obtained in Pomerania Province, together with a modified version of the method, which has been adjusted to the specificity of the British transport market and applied on the Milton Keynes–London route. The method used for the British market not only allows to assess which mode of transport is generally more competitive on a given route but also prompts which of them provides a better offer in case of early ticket purchase and which of them is more competitive for travellers buying their tickets right before departure. Importantly, the method is universal and can also be used to assess rail transport competitiveness on other routes.

Background

Regional passenger transport system consists of several modes of transport providing services of different characteristics. There is evident competition between private car and public transport, but also within public transport itself considering trains and buses as predominant service providers. In most cases travel pattern includes significant amount of commuting between central economic node of the region and its surroundings and other destinations. As Button [4] puts it: ‘Possibly the most important characteristics of transport is that it is not really demanded in its own right. (...) In other words, the demand for transport is a derived one but it's appropriate provision enables the benefits of a myriad of other, final benefits to be enjoyed'. One of the most important types of travel is those related to job and professional activity. They influence the economic situation of individuals and create an opportunity for other types of travels (leisure, education, etc.). The development of the region is affected by the share of its citizens included into core economic and social processes. Vickerman [5] states that changes in accessibility led to changes in the value of a region economic potential. According to Sampaio et al. [6], the capability to move easily in a territory is crucial for increasing the share of citizens in contemporary societies.

Kiel et al. [7] state that the transport system consists of three different markets, namely, the trip market, the transport market and the traffic market. Interactions between these markets and decisions made mark a dynamic character of this system based on the competition principle. The trip market consists of activities to be performed, with spatial and temporal distribution of locations where the activities could be performed. The output of this market is a set of trip patterns, an allocation of activities to location and times. The transport market is a response for trip patterns which demand for transport services of desired quality. Finally, the traffic market comprises the transport patterns which demand for infrastructure to accommodate the vehicles and services.

The accessibility to transport services is linked to its competitiveness and interactions between modes at the transport market. According to Caschili et al. [8], accessibility may be measured as a potential of opportunities, which can be reached from a given place at the cost of overcoming the friction associated with the movement through space and time. Competitiveness of selected transport modes and character of competition between them underlines the potential to improve accessibility across the region and for various groups of users. Geurs and van Wee [9] identified four types of components important in measuring accessibility: land use, transportation, temporal and individual. Direct references to competitiveness of regional transport services may be indicated there.

- The land use component consists of the amount and spatial distribution of opportunities supplied at each destination and the demand for these opportunities at origin destination. As a result, competition for these opportunities can begin favouring users and areas with better transport accessibility.
- The transportation component describes the transport system as disutility for an individual to cover distance between an origin and a destination using the specific transport mode. Factors included are the amount of time (travel, waiting, change, parking), costs (fixed and variable) and effort (reliability, level of comfort, accident risk, etc.). This disutility results in confrontation between the supply (i.e. public transport timetables and travel costs) and demand (volume of passengers' trips). Competitiveness of transport services can alter supply side to better facilitate the demand for services.

- The temporal component reflects the temporal constrains such as availability of opportunities at different times of the day and time available to participate in certain activities.
- The individual component reflects the needs (i.e. income, place of work), abilities (i.e. availability of travel modes) and opportunities (i.e. travel budget) of individuals. These characteristics influence a person's access to transport modes.

Assessment Method for the Polish Transport Market

Method Development

The competitiveness of various modes of transport can be assessed by checking which of them provides a better offer for passengers. According to Jackiewicz et al. [10], the primary criterion for assessing the quality of transport services is the realisation of demands declared by passengers. In order to check it, multiple assessment criteria can be taken into consideration. Previous works, such as Cieśla et al. [11], identified a number of possible criteria, including: travel duration, service frequency, number of changes, ticket price, availability, safety, punctuality, reliability, travel comfort, accessibility of passenger information, clarity of timetables, fare system simplicity and compatibility of routes and timetables with demand.

To assess and compare the competitiveness of different modes of transport on a given route, it is needed to develop a unified method which takes into consideration various criteria at a time. It has been decided to use arithmetic mean values of travel duration, ticket price, time between services (as a measure for service frequency) and number of changes. There are two reasons for choosing this set of criteria. Firstly, they are one of the essential criteria for passengers while planning a journey. Secondly, they have more objective nature, and it is possible to assess modes of transport in a numerical way with them. The list of selected criteria is:

- Average travel duration [minutes]
- Average ticket price [PLN]
- Average time between services [minutes]
- Average number of changes [–]

To adjust the importance of each of above elements, it has been decided to assign weights of importance to each of them. To do it, a survey has been conducted by the authors (in the year 2014) among people using the public transport in Pomerania Province at least once a month. 149 respondents have been asked, how important for them (in a scale of 1–10) is each of the four criteria when they choose a mode of transport for their travels. The survey has produced the following results:

- Travel duration—average rating of 8.68
- Ticket price—average rating of 8.12
- Service frequency—average rating of 8.28
- Number of changes—average rating of 7.79

Based on the results of the survey, appropriate weights have been assigned to all the elements. To do it, each average rating has been divided by the sum of all the average ratings. The results are as follows:

- Travel duration—weight of 0.2641
- Ticket price—weight of 0.2470
- Service frequency—weight of 0.2519
- Number of changes—weight of 0.2370

To calculate the Rail Transport Competitiveness Index (RTCI), the following equation, taking into account all the four elements with their weights of importance, has been proposed:

$$CI_{rail1} = 0.2641 \times \left(1 - \frac{TD_{rail}}{TD_{rail} + TD_{bus}}\right) + 0.2470 \times \left(1 - \frac{TP_{rail}}{TP_{rail} + TP_{bus}}\right) + 0.2519 \times \left(1 - \frac{TS_{rail}}{TS_{rail} + TS_{bus}}\right) + 0.2370 \times \left(1 - \frac{NC_{rail}}{NC_{rail} + NC_{bus}}\right)$$

where:

- CI_{rail1}—Rail Transport Competitiveness Index
- TD_{rail}—Average train travel duration [minutes]
- TD_{bus}—Average bus travel duration [minutes]
- TP_{rail}—Average train ticket price [PLN]
- TP_{bus}—Average bus ticket price [PLN]
- TS_{rail}—Average time between train services [minutes]
- TS_{bus}—Average time between bus services [minutes]
- NC_{rail}—Average number of changes between trains [–]
- NC_{bus}—Average number of changes between buses [–]

The Bus Transport Competitiveness Index (BTCI) will be calculated in the analogical way:

$$\mathrm{CI}_{\mathrm{bus1}} = 0.2641 \times \left(1 - \frac{\mathrm{TD}_{\mathrm{bus}}}{\mathrm{TD}_{\mathrm{rail}} + \mathrm{TD}_{\mathrm{bus}}}\right) + 0.2470 \times \left(1 - \frac{\mathrm{TP}_{\mathrm{bus}}}{\mathrm{TP}_{\mathrm{rail}} + \mathrm{TP}_{\mathrm{bus}}}\right) + 0.2519 \times \left(1 - \frac{\mathrm{TS}_{\mathrm{bus}}}{\mathrm{TS}_{\mathrm{rail}} + \mathrm{TS}_{\mathrm{bus}}}\right) + 0.2370 \times \left(1 - \frac{\mathrm{NC}_{\mathrm{bus}}}{\mathrm{NC}_{\mathrm{rail}} + \mathrm{NC}_{\mathrm{bus}}}\right)$$

where:

- CI_{bus1}—Bus Transport Competitiveness Index

If $NC_{rail} + NC_{bus} = 0$ on any analysed route (meaning that all train and bus services are direct), $NC_{rail}/(NC_{rail} + NC_{bus})$ and $NC_{bus}/(NC_{rail} + NC_{bus})$ will be substituted with the value of 0.5, giving equal number of points for the number of changes to both modes of transport.

Both equations will be used on each analysed route, providing two values within a range from 0 to 1. Index value closer to 0 indicates lower competitiveness of a given mode of transport, while value closer to 1 indicates higher competitiveness. The sum of both values for each route will be 1. In case of no rail or bus service running on daily basis on any analysed route, calculations will not be performed, and 0 value will be given automatically to this mode of transport, while 1 value will be given to the other mode of transport, providing it operates on daily basis.

Results in Pomerania Province

To assess the overall competitiveness of rail transport over bus transport in Pomerania Province, it has been decided to analyse regional train and bus services linking Tricity urban area (Gdańsk, Gdynia and Sopot) with 18 other major cities of the region (more than 20,000 inhabitants or status of a county town). For each of the 18 routes, the timetable data have been collected, and arithmetic mean values of travel duration, ticket price, number of changes and time between services have been calculated. Data have been collected for a weekday in June 2014. The scope of analysis has been narrowed down to regional and suburban services, as their fares are significantly lower which makes them a primary mode of transport inside the region. Because of the great similarity of offers for both travel directions on each route, data have been collected for one direction of travel only (from other cities to Tricity), but the numbers are representative for both. Table 1 presents average values for Malbork–Gdańsk route as an example of calculations which have been performed for all the 18 analysed routes.

With the average values presented above and the equations proposed earlier, RTCI and BTCI for the route have been calculated:

- Rail Transport Competitiveness Index: $CI_{rail1} = 0.70$
- Bus Transport Competitiveness Index: $CI_{bus1} = 0.30$

After calculating both indexes for all the 18 routes, a ranking of routes has been prepared, starting from the highest RTCI, as presented graphically in Fig. 1.

In order to provide a synthetic evaluation of the rail transport competitiveness in Pomerania Province, the RTCIs obtained have been divided into five step sizes, and a map of competitiveness has been prepared. As presented in Fig. 2, each route is highlighted with a different colour, indicating the level of rail transport competitiveness, with five levels being distinguished.

Table 1 Average values for Malbork–Gdańsk train and bus services

Malbork–Gdańsk	Rail transport	Bus transport
Average travel duration [h:min]	1:04	2:15
Average ticket price [PLN]	13.50	22.33
Average number of changes [–]	0.21	0.89
Average time between services [h:min]	1:16	2:40

Source: own elaboration

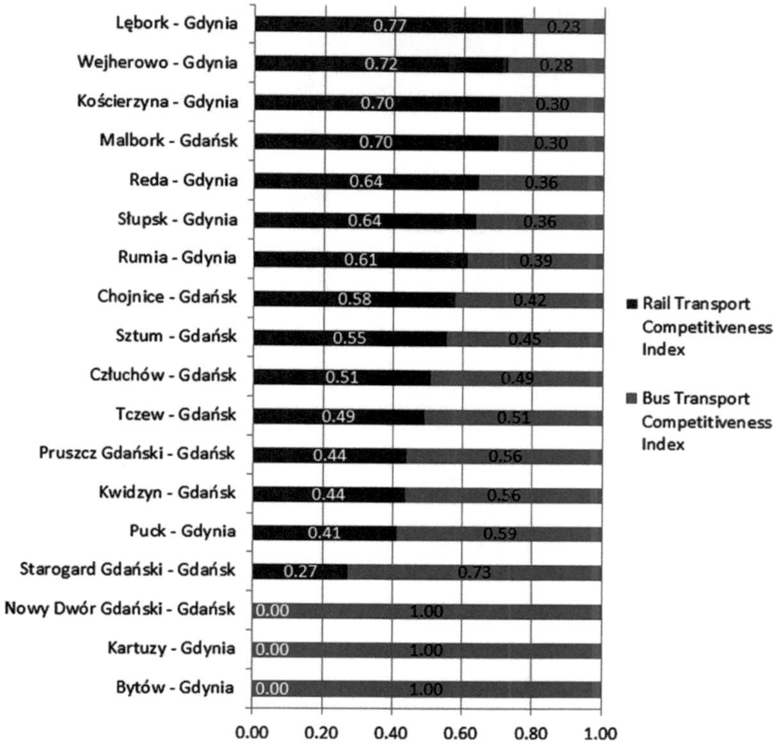

Fig. 1 Rail and bus transport competitiveness index values on the analysed routes (source: own elaboration)

As a next step, average RTCI and BTCI have been calculated, in order to assess the overall competitiveness of rail transport in the region:

- Average Rail Transport Competitiveness Index: $CI_{rail1} = 0.47$
- Average Bus Transport Competitiveness Index: $CI_{bus1} = 0.53$

With the above calculations, rail transport competitiveness assessment in Pomerania Province has been completed, providing information about each analysed route, as well as about the overall competitiveness level.

Discussion of Results

Several conclusions for the region can be drawn, based on the assessment performed:

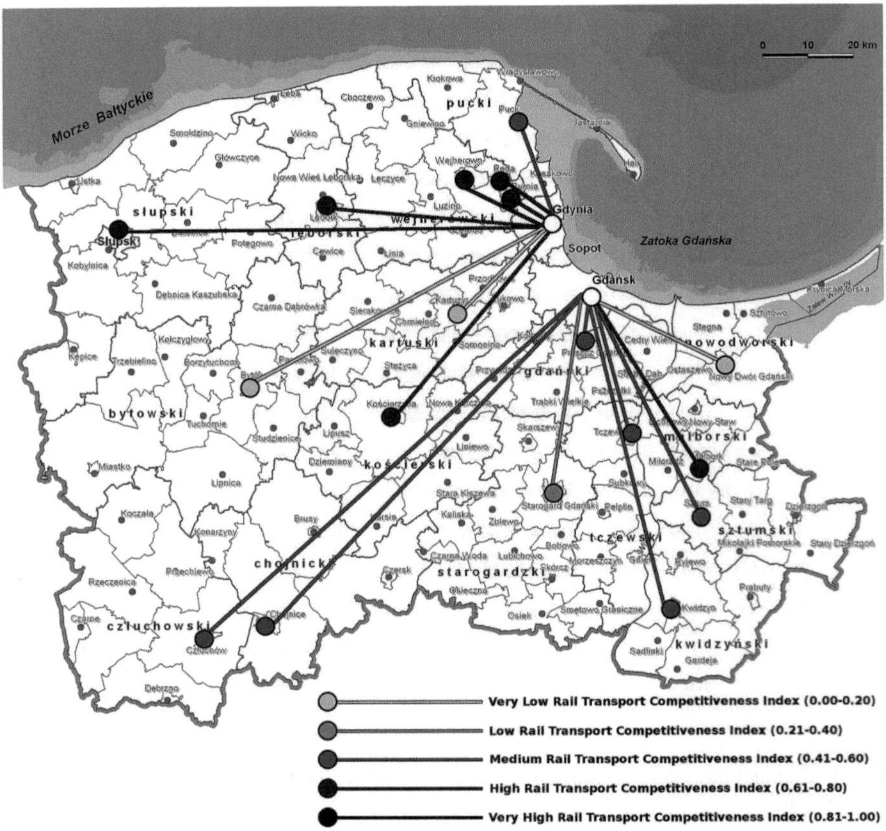

Fig. 2 Map of rail transport competitiveness in Pomerania Province (source: own elaboration based on the map sourced from http://midwig.pomorskie.eu/atlas_administracja_podzial.html)

- Rail transport is more competitive in the main railway corridor of the region, between Tricity and Słupsk, Lębork, Wejherowo, Reda, Rumia and Malbork and additionally on the Kościerzyna–Gdynia route.
- Exceptions from the above are Tczew–Gdańsk and Pruszcz Gdański–Gdańsk routes (which belong to the main railway corridor of the region), where the RTCI is on the medium level due to a strong competition of the bus transport.
- In most cases in the region, rail transport is more competitive on shorter distances, while bus transport is more competitive on longer distances.
- Rail transport competitiveness is generally higher in the northern part of the region.

Apart from the above, it can be noticed that despite the large variation of Rail and Bus Transport Competitiveness Indexes on all 18 analysed routes, their average values show very similar levels of overall competitiveness of both modes of transport in Pomerania Province. According to the results obtained, bus transport is a bit more competitive in the region but the difference is not significant.

Assessment Method for the British Transport Market

Method Modifications

The method developed for the Polish regional transport market is generally universal and can be used for other routes and markets too. However, they must fulfil some requirements, e.g. the prices have to be constant and cannot vary in time. It has been decided that to apply the method for the British transport market, it must be slightly modified. Milton Keynes–London route has been chosen to test the modified method and assess the rail transport competitiveness. The following modifications to the method have been made to adjust it to the requirements of the British transport market and Milton Keynes–London route in particular:

- All train and bus services have been taken into account instead of distinguishing regional and long-distance services. This is because there is no significant difference between regional and long-distance service fares, with some long-distance service fares being even lower.
- The number of changes has been removed from the equation, as all the train and bus services between Milton Keynes and London are direct. This modification has been made specifically for the analysed route, and it does not necessarily apply to the other British routes.
- New survey has been conducted among passengers travelling between Milton Keynes and London, in order to assign new weights of importance to all the criteria, suitable for the analysed route.
- Prices of tickets have been collected twice, as a system of dynamic fares is in use on the analysed route and some of the prices vary in time. Data have been first collected 1 week before the date of travel and then on the day of travel, before departure times.

The survey has been conducted by the authors (in the year 2015), and 23 respondents have answered how important for them (in a scale of 1–10) is each of the three elements when they choose a mode of transport for their travels. The survey has produced the following results:

- Travel duration—average rating of 8.30
- Ticket price—average rating of 7.61
- Service frequency—average rating of 8.26

Based on the results of the survey, appropriate weights have been assigned to all the elements. To do it, each average rating has been divided by the sum of all the average ratings. The results are as follows:

- Travel duration—weight of 0.3434
- Ticket price—weight of 0.3149
- Service frequency—weight of 0.3417

Based on the above, a modified process has been proposed to calculate the RTCI for Milton Keynes–London route. The process consists of three steps this time:

1. Calculations with early ticket price data, collected 1 week before the travel date
2. Calculations with late ticket price data, collected on the day of travel
3. Calculations of the arithmetic mean of two values calculated in preceding steps

A modified version of the earlier equations has been proposed to calculate RTCI and BTCI, considering new weights of importance and lack of number of changes in the calculations:

$$CI_{rail2} = 0.3434 \times \left(1 - \frac{TD_{rail}}{TD_{rail} + TD_{bus}}\right) + 0.3149 \times \left(1 - \frac{TP_{rail}}{TP_{rail} + TP_{bus}}\right) + 0.3417 \times \left(1 - \frac{TS_{rail}}{TS_{rail} + TS_{bus}}\right)$$

$$CI_{bus2} = 0.3434 \times \left(1 - \frac{TD_{bus}}{TD_{rail} + TD_{bus}}\right) + 0.3149 \times \left(1 - \frac{TP_{bus}}{TP_{rail} + TP_{bus}}\right) + 0.3417 \times \left(1 - \frac{TS_{bus}}{TS_{rail} + TS_{bus}}\right)$$

where:

- CI_{rail2}—Rail Transport Competitiveness Index
- CI_{bus2}—Bus Transport Competitiveness Index
- TD_{rail}—Average train travel duration [minutes]
- TD_{bus}—Average bus travel duration [minutes]
- TP_{rail}—Average train ticket price [GBP]
- TP_{bus}—Average bus ticket price [GBP]
- TS_{rail}—Average time between train services [minutes]
- TS_{bus}—Average time between bus services [minutes]

Both equations are used at the first and second step of the process. The values obtained in the first step will be called Early Purchase RTCI and Early Purchase BTCI, respectively. Similarly, the values obtained in the second step will be called Late Purchase RTCI and Late Purchase BTCI, respectively. The arithmetic means calculated in the third step will be called Overall RTCI and Overall BTCI.

Results on Milton Keynes–London Route

The modified method has been applied to the selected route. Data have been collected for a weekday in February 2015. Same as previously, data have been collected for one direction of travel only (from Milton Keynes to London), but the

Table 2 Average values for Milton Keynes–London train and bus services

Milton Keynes–London	Rail transport	Bus transport
Average travel duration [h:min]	0:45	1:46
Average early ticket price [GBP]	12.60	12.08
Average late ticket price [GBP]	15.56	12.08
Average time between services [h:min]	0:10	1:25

Source: own elaboration

numbers are representative for both because of the great similarity of offers for both directions. After collecting all the necessary data, average values of travel duration, early ticket price, late ticket price and time between services have been calculated, as presented in Table 2.

With the average values presented above and the equations proposed earlier, RTCI and BTCI for the route have been calculated, first for early purchases:

- Early Purchase Rail Transport Competitiveness Index: $CI_{rail2} = 0.70$
- Early Purchase Bus Transport Competitiveness Index: $CI_{bus2} = 0.30$

In the second step, RTCI and BTCI have been calculated for late purchases:

- Late Purchase Rail Transport Competitiveness Index: $CI_{rail2} = 0.68$
- Late Purchase Bus Transport Competitiveness Index: $CI_{bus2} = 0.32$

Finally, the overall indexes have been calculated:

- Overall Rail Transport Competitiveness Index: $CI_{rail2} = 0.69$
- Overall Bus Transport Competitiveness Index: $CI_{bus2} = 0.31$

With the above calculations, rail transport competitiveness assessment on Milton Keynes–London route has been completed, providing information about competitiveness for passengers buying tickets 1 week in advance, about competitiveness for passengers buying tickets on departure day as well as about the overall competitiveness level.

Discussion of Results

The assessment performed shows a significant competitive advantage of rail transport on the analysed route. It also shows that the rail transport is more competitive for passengers buying tickets in advance. While buying tickets on the departure day, the competitive advantage is smaller but remains significant. Based on the data collected, it can also be noticed that the average prices of tickets bought in advance are very similar for both modes of transport, while the average prices of tickets bought late are significantly different, with rail tickets being more expensive. However, the results show that the advantage of bus transport in the area of ticket prices is not sufficiently large to achieve an overall competitive advantage over rail transport on this route.

Conclusions and Future Work

The method has proved to be useful in assessing rail transport competitiveness on different markets, having their own rules and specificities. Despite the fact of being able to use the method on various markets, the need for its modifications in specific circumstances has been found out. The local requirements have always to be taken into account, in order to make the method reliable in providing accurate results, as the example of dynamic fare system shows. The results obtained in both countries show a big variety of rail transport competitiveness levels on different routes. Significant differences in RTCI and BTCI show the ability of the developed method to distinguish various competitiveness levels in a clear numeric way.

As a promising modification to the method, the way of assessing the importance of various criteria for choosing a mode of transport can be improved. More comprehensive studies would help to assign weights of importance more accurately, both in Poland and the United Kingdom, making the method more precise. Improvements can be made both in terms of survey methodology and sample size. Changes in methodology may possibly help differentiating the weights of importance more, making them more accurate. However, only a limited impact on the overall results is expected. It should be noticed that the results of survey for the British market are accurate for one particular route, as the respondents could take into consideration its own characteristics. Similarly, the way of assigning weights of importance for the Polish market could be improved by differentiating them for each route, as they may vary depending on a local specificity.

The method has been found to be promising not only in terms of the current competitiveness assessment but also in terms of simulation of competitiveness after transport upgrade projects being performed. Its use for simulation can prompt transport companies and local authorities on the competitiveness levels of different transport modes after prospective upgrades and changes in the offer for passengers. This can also be useful for decision making, prompting what level of prices or journey time is needed to achieve the competitive advantage.

References

1. Smoliński, S., 2007. Determinanty konkurencyjności na kolejowym rynku przewozów pasażerskich. In: Bernat, T., 2007. Przedsiębiorstwo i państwo - wybrane problemy konkurencyjności. Katedra Mikroekonomii Uniwersytetu Szczecińskiego, Szczecin.
2. Rosik, P., Stępniak, M., Komornicki, T., 2015. The decade of the big push to roads in Poland: Impact on improvement in accessibility and territorial cohesion from a policy perspective. Transp. Policy 37, 134–146.
3. Jurewicz, P., 2014. Analiza konkurencyjności kolei wobec innych rodzajów transportu zbiorowego na terenie województwa pomorskiego. Wydział Inżynierii Lądowej i Środowiska, Politechnika Gdańska, Gdańsk.
4. Button, K., 2010. Transport Economics. Edward Elgar Publishing Ltd.

5. Vickerman, R., 1995. Location, accessibility and regional development: the appraisal of trans-European networks. Transp. Policy 2, 225–234
6. Sampaio, B.R., Neto, O.L., Sampaio, Y., 2008. Efficiency analysis of public transport systems: Lessons for institutional planning. Transp. Res. Part A Policy Pract. 42, 445–454
7. Kiel, J., Smith, R., Ubbels, B., 2014. The Impact of Transport Investments on Competitiveness. Transp. Res. Procedia 1, 77–88.
8. Caschili, S., De Montis, A., Trogu, D., 2014. Accessibility and rurality indicators for regional development. Comput. Environ. Urban Syst. 49, 98–114.
9. Geurs, K.T., van Wee, B., 2004. Accessibility evaluation of land-use and transport strategies: Review and research directions. J. Transp. Geogr. 12, 127–140.
10. Jackiewicz, J., Czech, P., Barcik, J., 2010. Standardy jakości usług w komunikacji miejskiej – część 1. Zeszyty Naukowe Politechniki Śląskiej, seria Transport, z. 67, 2010.
11. Cieśla, M., Gąska, D., Opasiak, T., 2010. Preferencje pasażerów transportu zbiorowego na terenie Aglomeracji Śląskiej. Transport Miejski i Regionalny 9/2010.

The Impact of Public Procurement System on Concentration of the Rail Freight Market in Poland

Grzegorz Krawczyk

Abstract The aim of this chapter is to examine the relations between public transport service procurement and the level of concentration of rail freight transport market. This chapter analyses the demand reported by the public sector for the services of rail freight transport in Poland. Procurement notices were characterised with regard to type of load, reported demand for transport activity volume and applied tendering procedure. This chapter presents the share of demand reported in public tenders against the total market demand. Moreover, in order to identify the market concentration level, measurement was performed on the basis of discrete and accumulated measures.

The analysis of the share of the volume of transport services purchased in tenders led to the conclusion that it could play a part in shaping market behaviours of units in the highly concentrated rail freight transport market. Demand executed through public procurement procedure may also have an impact on the market structure.

Introduction

Transport is one of the most regulated sectors of national economy. The transport liberalisation process in Poland is very diverse, depending on the branch and characteristics of a particular market. The rail freight transport market in Poland has gone through a long transformation process. In European rankings, such as Rail Liberalisation Index 2011, Poland is placed in the group of countries with a high liberalisation index. However, the Polish rail transport market is still characterised by a relatively high level of concentration. The concentrated market structure may have an impact on behaviours of companies. In the open economy model, apart from the demand reported by the private sector, we can also distinguish demand generated by institutions and companies governed by the public procurement law. Guided by the principle of optimisation and reasonableness of public expenditure, such units are obliged to apply tendering procedures in accordance with the existing legal system. A significant volume of services, as well as a long term of contractual

G. Krawczyk (✉)
Chair of Transport, Economic University of Katowice, Katowice, Poland
e-mail: grzegorz.krawczyk@ue.katowice.pl

M. Bąk (ed.), *Transport Development Challenges in the Twenty-First Century*, Springer Proceedings in Business and Economics,
DOI 10.1007/978-3-319-26848-4_4

obligation, may have impact on shaping the intensity of competition and market behaviours of units in a particular market.

The aim of this chapter is to analyse the impact of public procurement on the level of concentration. Additionally, public tenders for rail freight transport (in the territory of Poland) announced in the Official Journals of the European Union and published on Tenders Electronic Daily were identified and the corresponding procurement notices were characterised with regard to the type of load, reported demand for transport work volume, and results of tender proceedings.

Rail Freight Transport in Poland

Public sector is involved in transport processes on many levels [1]:

- From the supply perspective, as the owner of state and municipal enterprises handling transport
- From the demand perspective, as the customer requesting for transport of passengers (e.g. inhabitants of a commune—public transport), or freight (e.g. state-owned enterprises in the energy industry)
- As a market regulatory agency

Considering the role of the state as a customer, we should bear in mind the clarity and transparency of financial management, which is necessary for the public sector. In consequence, public procurement is executed through tendering proceedings (according to various formulas and value thresholds), which is intended to enable the highest possible number of market players to compete. The purpose of competition between companies is to achieve the lowest possible price for the expected quality of services.

Freight transport in Poland is very diverse with regard to the involvement of particular modes of transport. In terms of total transported weight of freight, the market constantly increases; however, this development is uneven, with a significantly strong growth of importance of vehicle transport. Due to the lack of significant investment in construction of new infrastructure, or modernisation of the already existing railway connections over the last decade, rail transport remains stagnant.

The Polish rail freight transport market is determined by the following factors: quality of infrastructure (which has the impact, e.g. on the average commercial speed), and rolling stock capacity and transport demand structure. The consequence of this specific character is the dominating position of raw materials in the subject-matter structure of freight transport. Transport of raw materials has the largest share in the total volume of transported freight. The dominating type of load in 2010–2013 was hard coal, which constituted more than 40 % of the total volume of transported freight. Other significant freight groups included metal ores (25 %), stone, sand, gravel, and clay (over 10 %) [2–5].

According to LIB (Liberalisation Rail Index) [6], which describes the degree of implementation of European law in the field of rail transport, the Polish market has been at the stage of advanced liberalisation since 2011. Despite clear market entry

Table 1 Measurement of concentration in the rail freight transport market

Index	2011	2012	2013
HHI—according to the weight of transported freight	3226	3061	2858
CR (4)—according to the weight of transported freight	73	82	79
HHI—according to transport activity	4153	3843	3668
CR (4)—according to transport activity	84	82	80

Own study

procedures, the analysis of concentration level indicates its monopolistic character. The concentrated nature of the rail freight transport market should be taken into consideration during the analysis of its structure of entities. The dominating entity in terms of the market share is PKP Cargo.

From the perspective of the examination of behaviours of entities acting in the transport market, as well as assessment of the intensity of competition between them, the key issue becomes the measurement of market concentration, understood as the function of the number of companies and shares in the market owned by them. Market concentration is the index measuring the degree to which a relatively small number of companies have a significant share in the market with regard to the volume of provided transport services.

The theory of statistics developed many concentration measures, which could be divided, e.g., into accumulated and discrete types. Discrete measures include several largest selected entities within the scope of their analysis, whereas accumulated measures are based on the participation of all selected market participants [7]. Due to the universal character and clarity of calculations, Herfindahl–Hirschman Index, widely used in the practice of anti-monopoly offices, was applied in order to express the level of market concentration [8]. In mathematical expression, it is the sum of squares of shares of all market players [9]. Additionally, the values of concentration index were determined for four entities (Table 1).

When interpreting the values of market concentration indices, one may conclude that this market is characterised by a high level of concentration. A gradual decline of concentration is noticeable in 2011–2013, mainly with regard to the weight of transported freight. The functioning of one entity or a narrow group of dominating entities in the market may have many negative effects for shaping the level of competition. The dominating entity may use the predatory price mechanism and gradually limit the market share of smaller entrepreneurs. The economic effects of such hypothetical situation are unfavourable in long-term perspective and create the risk of price increase.

Characteristics of Public Procurement for Rail Transport Services

In order to identify the demand volume requested by the entities obliged to apply tendering procedures, the publications of the Supplement to the Official Journal of the European Union were used. It was assumed that the demand for rail freight transport services would be requested mainly by entities from the energy and extraction sectors. If the State Treasury is a shareholder, such entities are obliged to apply the public procurement provisions and are referred to as sector ordering parties. In consequence, until 1 January 2014 the amount of the so-called EU threshold (above which an announcement should be placed in the Supplement) was estimated as EUR 400,000. Due to the cost-consuming nature of rail transport, as well as the expected significant volume of purchase, it may be assumed that all important contracts in the Polish market are higher in value than the above-mentioned amount and therefore have been placed in the Supplement to the Official Journal of the European Union.

Due to the data availability, the period of 5 full years, i.e. from 2010 to 2014 inclusive, was adopted for the analysis of tenders for rail freight transport services. Given the complexity of tendering procedures, the possibility of submitting appeals as well as time required for their analysis by Krajowa Izba Odwoławcza (National Appeal Chamber), only the data included in the contract awarding announcements which had been published and occurred in the analysed period were taken into consideration.

In the analysed period, 67 public procurement contracts were awarded. The public procurement tenders were mostly announced by one entity, and group tenders were rare. The agreements with the contractors were usually concluded for 1-year cooperation periods, which resulted in the cyclicality of certain public procurements. The value of rail transport services ordered by the entities obliged to apply Ustawa Zamówień Publicznych (UZP, Public Procurement Law) in 2010–2014 amounted to over PLN 2400 million. The actually achieved prices of contracted services were overall lower by approximately 21.5 % than the initial estimates of the ordering parties. Given the partial recurrence of purchases in the following years, this may indicate aggressive competition between the tenderers (Table 2).

In the group of 67 contract award notices, in most cases the volume of load weight and the load type were precisely determined. In the case of 13 notices, the public procurement value was revealed, whereas the load weight was not revealed. In all cases, public demand for rail freight transport was related to mass freight, and in particular to the transport of raw materials.

In the analysed period, the total amount of over 105 million tons of freight was contracted—mainly of coal (almost 98 % of the entire volume). In the case of contracts for a period longer than 1 year, the provided documents do not contain precise data concerning the distribution of transport activity between particular years, and

Table 2 Demand requested in the form of public procurements

Year	Number of tenders	Volume of public procurements (in tons)
2010	13	6,303,100
2011	18	17,776,650
2012	10	27,265,916
2013	15	29,365,783
2014	11	24,635,333

Own study

therefore the values adopted for the purpose of the research were based on arithmetic average values, with the assumption that transport activity distribution is proportional to the length of contract.

The analysis of procurement notices allows to conclude that the entities ordering rail freight transport services within the framework of tendering proceedings are companies in extraction and energy sectors. A number of dominating entities can be distinguished with regard to the requested volume of transport load freight:

- Polska Grupa Energetyczna (the entire capital group): approximately 35.4 %
- Elektrownia Kozienice (Enea Wytwarzanie): approximately 25.1 %
- Grupa Tauron: approximately 26.2 %

It shall be emphasized that not all contract award notices contain data on the total weight of raw materials. For example, the public procurements announced by the mining companies: Kompania Węglowa and Katowicki Holding Węglowy were specified as "provision of services related to rail transport of coal from coal mines of Kompania Węglowa S.A. to the recipients of coal". The accuracy of the provisions of the content of tender documents results from the applied public procurement procedure.

Open procedure was the dominating form of proceedings in terms of the number (61.1 %) and value of contracts (74.4 %). In all cases of applying open procedure, the only bid selection criterion was the lowest price. The least transparent form is utility procurement, awarded through public procurement procedure without prior notice publication. In the analysed period, it constitutes approximately 18 % of the total number of all public procurements and amounts to approximately PLN 170 million (7.03 of the total amount). The type of procedure may have an impact on the intensity of competition between carriers. Transparency of tendering proceedings may enable market verification of prices between the main market players.

The Role of Public Procurement in Shaping Market Competition

The main aims of applying the public procurement procedure are to rationalize expenditure and to enable new companies, which do not have close business relations with the ordering party, to be awarded a contract. Tendering procedure supports the increase of competition and improves the understanding of prices by the ordering party. In the case of public tenders, the important elements are appropriate terms of reference and determination of the bid selection criterion. Preparation of terms of reference for transport of raw materials is not a complicated task. Simple and universal transport technology helps determine the bid selection criterion only on the basis of the lowest price. As a consequence, such situation may induce market participants to lower their prices deliberately in order to be awarded a contract (Table 3).

The table contains a summary of contractors awarded in the analysed tendering proceedings in 2010–2014, with regard to the freight volume and the final value of contract. The biggest number of contracts were awarded to the main entity in the

Table 3 Summary of tender results in 2011–2014

Contractor	Weight volume (tons)	Amount estimated by the ordering party (PLN)	Price (PLN)	Percentage according to mass (%)	Percentage according to price (%)
PKP Cargo	84,052,920	2,333,187,784	1,703,522,551	75.96	71, 18
DB Schenker Rail Polska	20,147,830	508,493,328	532,684,614	18.21	22.26
CTL Logistic + Rail Polska	2,473,000	43,973,166	35,461,496	2.23	1.48
Freightliner PL	485,000	17,999,100	15,975,000	0.44	0.67
Pol-Miedź Trans	2,805,000	97,349,000	72,118,800	2.53	3.01
CTL Logistic	625,800	32,582,575	31,962,700	0.57	1.34
Cemet	64,500	1,845,600	1,632,570	0.06	0.07

Own study

market of mass rail transport, i.e. PKP Cargo. The company contracted services for the transport of over 84 million tons of freight (mainly coal), which amounted to almost 76 % of tonnage reported in all proceedings. This allowed for achieving the income in the amount of over PLN 1700 million. The second entity, according to the criterion of awarded transport, was DB Schenker Rail Polska, with the share of 18.21 % in the total number of procurements and 22.26 % according to their market value. The overall share of other entities in the public procurement market was at the level of less than 6 % of the total public procurement value. In all analysed contract award notices, the only selection criterion was the lowest price. Therefore, the performed analysis allows to conclude that PKP Cargo is the most active entity in the market of transport requested by public entities. It shall be noticed that within the scale of all analysed procurements, the final contract prices are approximately 22 % lower than estimated by the ordering party. In the case of contracts awarded to PKP Cargo, the difference amounts to 27 %. On the one hand, given the cyclicality of part of the procurements, the estimates of ordering parties may be regarded as the determinant of market prices, on the other hand, they may be calculated at a high level of generality. Despite the above-mentioned doubts, such significant discrepancies between the estimates and the prices reported by PKP Cargo have to be taken into consideration. Based on this, it may be concluded that PKP Cargo, as the largest company in the Polish rail freight transport market, pursues an aggressive pricing policy in the public procurement market.

The analysis of the volume of demand submitted by the public finance sector (companies with participation of the State Treasury) in 2010-2013 (data for 2014 are not available yet) allows to conclude that it has a potential impact on shaping market relations (Table 4). Due to the concentrated market structure, winning contracts through tenders may change the balance of forces between the carriers. The structure of types of transported loads and demand for coal transport will not change over the next couple of years due to the strong dependence of industry on this raw material. This situation may also be used by the dominating entity, which may strengthen its

Table 4 The share of public sector demand in the rail transport market according to load weight

Year	Total transported load weight (tons)	Demand reported by the public sector (tons)	Share (%)
2010	235,500,000	6,303,100	2.68
2011	249,300,000	17,776,650	7.13
2012	231,300,000	27,265,916	11.79
2013	233,200,000	29,365,783	12.59

Own study

position in the long-term perspective through gradual elimination of its competition by applying lowered prices in tenders. The calculations presented in this chapter demonstrate a gradual decrease of market concentration, as well as limitation of the significance of the largest carriers. The volume of public procurement is large enough to have an impact on the structure of the market, and the selection of a bid based on the lowest price criterion may result in applying predatory prices. This is confirmed by numerous appeals to Krajowa Izba Odwoławcza (National Appeal Chamber), submitted by tender participants, who report the application of grossly low prices.

Summary

The analysis of the data related to the reported demand of the public sector (energy and extraction companies with the participation of the State Treasury) allowed to determine the ordered volume of transport, type of load, applied tendering procedures and value of contracts. In the adopted analysis period (2010–2014), 67 contract award notices were identified. Among them, open tender procedure, which ensures the highest transparency of proceeding, was the dominating one. In the vast majority of cases, the transported raw material was hard coal. The total value of concluded agreements amounted to over PLN 2400 million, whereas the contracted load weight volume exceeded 105 million of tons. The most active carrier in the field of winning contracts through tenders was PKP Cargo, which achieved the share of over 70 % in the analysed contracts.

Taking into consideration the structure of the rail transport market in Poland, its concentrated character, as well as the dominating role of PKP Cargo, should be emphasised. The share of procurements announced by energy and extraction companies in the structure of the entire market ranged between a few and several per cent and could have had an impact on shaping the intensity of competition and market behaviours of carriers. In the analysed period, the gradual decrease of concentration indices in the market of rail freight transport in Poland can be observed. The main aim of this chapter was to study the relations between public transport service procurement and the level of the concentration on the rail transport market. The resolution is the following: in the years 2010–2013 a gradual decrease of the concentration in the rail freight transport market was observed. Thus the relationship between procurement and the level of market concentration is inversely proportional. However the structure of tender results is characterised by higher concentration than the entire market, whereas the volume of transport activity won in this manner may have an impact on changes in the market structure in the long run (for example: the increase of concentration of rail transport market).

References

1. Transport. W. Rydzykowski, K. Wojewódzka-Król (eds), p. 38, Wydawnictwo Naukowe PWN, Warszawa (1997).
2. Funkcjonowanie rynku transport kolejowego w Polsce w 2010 roku, Office of Rail Transport, Warszawa, (2011).
3. Funkcjonowanie rynku transport kolejowego w Polsce w 2011 roku, Office of Rail Transport, Warszawa, (2012).
4. Ocena funckjonowania rynku transportu kolejowego i stanu bezpieczeństwa ruchu kolejowego w 2012 roku, Office of Rail Transport, Warszawa, (2013).
5. Ocena funckjonowania rynku transportu kolejowego i stanu bezpieczeństwa ruchu kolejowego w 2013 roku, Office of Rail Transport, Warszawa, (2014).
6. Rail Liberalisation Index 2011, IBM Global Business Services.
7. Bailey D., Boyle S.E.: The optimal measures of concentration. In: Journal of the American Statistical Association, Vol. 66, No. 336, pp. 702-703 (1971).
8. Cirba S., Ginevicius R.: Additive measurement of market concentration. In: Journal of Business Economics and Management. No. 10 (3). pp. 192 (2009).
9. Hall M., Tideman N.: Measures of concentration. In: Journal of the American Statistical Association. Vol. 62, No. 317, pp. 163-164 (1967)

Local Government, The Hostage or Playmaker in Competition Between Modes of Transport: Examples from Local Freight Transport in Lower Silesia

Krzysztof Lewandowski

Abstract The chapter presents the relationship between rights and responsibilities of local government in the planning of the transport system in the subordinate region, and competition between transport modes, especially land transport, road or rail. Rights and responsibilities of local government in the planning of the transport system in the subordinate region stem from laws on local government. Competition between modes of land transport derived from the principles of the Law on freedom of economic activity and the progressive liberalization of access to the transport market of the European Union countries. This gives rise to conflicts of service operators in the region. Local government may then have the choice of how to reconcile sustainable transport strategies in their territory and the desire to shape the transport of environmentally friendly behavior by generating instruments to internalize external costs of transport. This chapter provides examples of the decisions in these cases of the local government from Lower Silesia in Poland.

Keywords Local government • Sustainable transport strategy

Introduction

Local government is a form of public administration which exists as the lowest tier of administration within a given state. Local governments act within powers delegated to them by legislation and directives from the higher level of government. The chapter provides a discussion on responsibilities of the local government in the area of transport in Poland.

K. Lewandowski (✉)
Department of Maintenance and Operation of Logistic, Transportation and Hydraulic Systems, Wrocław University of Technology, Wrocław, Poland
e-mail: krzysztof.lewandowski@pwr.edu.pl

M. Bąk (ed.), *Transport Development Challenges in the Twenty-First Century*, Springer Proceedings in Business and Economics,
DOI 10.1007/978-3-319-26848-4_5

Review of Legal Regulations

The constitution of the Republic of Poland (art. 164) specifies that the community is the basic unit of local government [1]. It regulates in detail the Act of community, 1990 [2]. Other local government units are: the county ("powiat") [3] and the province ("wojwództwo") [4]. Accordingly, it is governed by Law acts—(Act of county, 1998) Dz.U.1998 No. 91, item. 578. The Act of 5 June 1998. Of county government and (Act of province, 1998) Dz.U. 1998 No. 91, item. 576. The Act of 5 June 1998. Provincial self-government. Each of them has the right of regulation in the subject of transport (Table 1).

As can be observed, all forms of local government are responsible for regulations of the road transport (community and county) and also other forms of transport, such as rail, water, and air—within the province.

The report on future plans for local administrative territory suggests that the subject of transport is connected with the road transport a number of times.

When Poland entered the European Union in 2004, many local administrative territories started to decide to analyze the significance of rail transport for this area. One can see that the discussion covered mainly the passenger and freight transport.

Competition Between Modes of Transport

After the collapse of the previous political system—communism in 1989, where all decisions were based on the opinion of state government, all of the economic conditions changed. In 1990, Act of community became the basic legal act for the local government. In this act is described the area of responsibility for the community. Economy changed rules for all carriers. Many communities started to regulate the local public transport. Because at this time all rail infrastructure was owned by the state carrier—PKP Polskie Koleje Państwowe, Polish State Railways, no local government could decide on rail transport. In Poland the road transport is the main carrier of goods and passengers [18, 19].

After developing the road infrastructure, there was an increase in the number of licenses for freight, passenger, and own needs transport (Fig. 1).

This situation influenced the opinion of administration of local government. In many cases in debates representatives decide to spend funds on roads.

The sources of income for local government are the second element of competition [18] between the modes of transport [19]. Each local government has different sources of income (Table. 2).

The analysis of Table 2 suggests that each local government decides about road transport. This is the basis for favoring the road transport.

Table 1 Responsibility of local government

Local government	Art. 7. Community [2]	Art. 4. 1. County [3]	Province [4]
Responsibility of	1) spatial management, real estate, environmental, and nature and water management; 2) municipal roads, streets, bridges, squares, and traffic organization; 4) the local public transport; 11) markets and exhibition halls; 14) public order and safety, and fire protection incandescent and flood control, including equipment and maintenance of municipal flood storage;	6) public transport and public roads; 13) environmental protection and nature conservation; 15) public order and security of citizens;	Art. 11. 1. The province government determines the development strategy of the region, taking into account in particular the following objectives: 2) the stimulation of economic activity; 3) improvement of economic competitiveness and innovation of the province; 2. the regional government establishes the policy of development for the region, which comprises: 1) creating conditions for economic development, including the creation of the labor market; 2) the maintenance and development of social and technical infrastructure of the importance of provincial; Art. 14. 1. The province government shall perform the duties of provincial laws, in particular: 7) spatial development; 8) environmental protection; 10) public transport and public roads;

Own work based on [2–4]

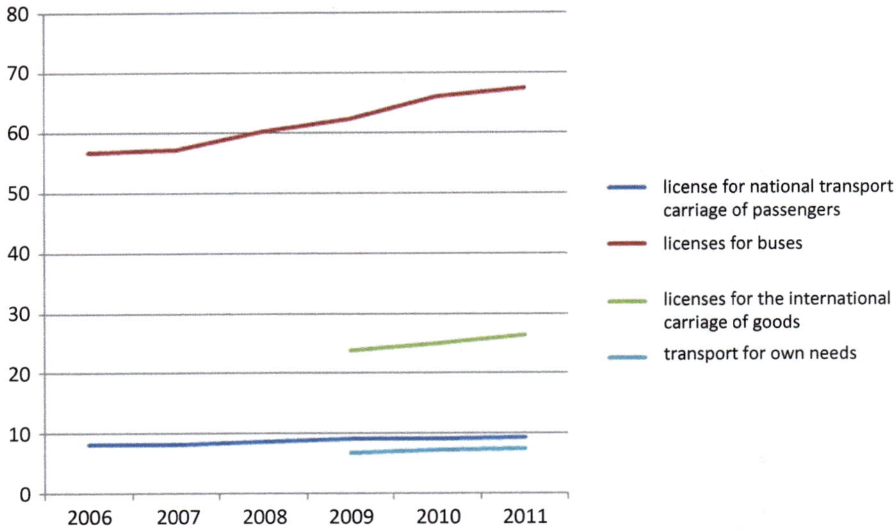

Fig. 1 Number of licenses for freight, passenger, and own needs transport (based on Refs. [5, 6])

Local Economy as Source of Sustainable Transport Strategies

In order to increase the economy after the change of political system, a law act concerning special economic zones was introduced in 1994 (pol. Ustawa z dnia 20 października 1994 r. o specjalnych strefach ekonomicznych. Dz.U. 1994 nr 123 poz. 600) [7].

Special economic zones were established to attract direct foreign investments. Today there are 14 special economic zones in Poland (Fig. 2).

Sources of income for local government are linked with different modes of economy. Distribution of major industrial companies in Poland in 2008 is shown in (Fig. 3).

Each new investment in the region generates new jobs and for local government this is a possibility to get new income.

New investments require from the local government some investment in infrastructure. Because legal regulations get no possibility for creation and regulation of rail infrastructure, local governments decided to build or modernize roads a number of times.

Sometimes new investors advertise themselves as ecological companies and they would decide to build a railway sidings. Local governments give special reduction of taxes from ground. After nearly 15 years we can see these railway sidings with trees that are 3 m high. In this chapter I do not present names and localizations of such companies because I do not wish to find myself in court.

Other positive situations are if the local government decided to use a possibility which are in the new law acts.

Table 2 Sources of income of local government (Act of community, 1990); (Act of county, 1998), (Act of province, 1998)

1	Art. 4 Community	Art. 5. County	Art. 6 Province
Sources of income	1. The sources of income of the community are: 1) receipts from taxes: a) from the property, b) agricultural, c) forest, d) on means of transport, e) personal income, paid in the form of a tax card, f) (repealed) g) on inheritance and gifts, h) on civil law transactions; 2) the amount of fees: a) duty, b) the fair, c) the local fees, charges of the spas and on possession of the dogs, d) (repealed) e) operational—in part referred to in the Act of February 4, 1994.—Geological and Mining Law (Dz. U. of 2005. No. 228, item. 1947, as amended. d.), f) other income which of community to be paid on the basis of separate regulations; 3) the income received as communal budget units and payments from municipal budgetary establishments; 4) income from the assets of community 5) inheritances, bequests, and donations to the community; 6) income from penalties and fines specified in separate regulations;	1. The sources of own revenues of the county are: 1) the amount of fees which the county income, paid on the basis of separate regulations; 2) the income received by the county budget units and payments from the county budget establishments; 3) income from the property of the county; 4) inheritances, bequests, and donations for the county; 5) income from penalties and fines specified in separate regulations; 6) 5.0 % of the revenue generated to the state budget in connection with the execution of the tasks of government administration and other tasks assigned to laws, unless separate regulations decide otherwise; 7) interest on loans granted by the county, unless separate regulations decide otherwise; 8) interest on late charges, which are provided to the county income; 9) interest on funds held in bank accounts of the county, unless separate regulations decide otherwise;	1. The sources of income of their own province are: 1) the income received by the provincial budget units and payments from provincial budgetary establishments; 2) income from assets of the region; 3) inheritances, bequests, and donations to province; 4) income from penalties and fines specified in separate regulations; 5) 5.0 % of the revenue generated to the state budget in connection with the execution of the tasks of government administration and other tasks assigned to laws, unless separate regulations decide otherwise; 6) interest on loans made by the province unless separate regulations decide otherwise; 7) interest on the overdue receivables representing proceeds transferred province;

continued

Table 2 (continued)

1	Art. 4 Community	Art. 5. County	Art. 6 Province
	7) 5.0 % of the revenue generated to the state budget in connection with the execution of the tasks of government administration and other tasks assigned to laws, unless separate regulations decide otherwise; 8) interest on loans granted by the municipality, unless separate regulations decide otherwise; 9) interest on the overdue receivables representing proceedings transferred of community 10) interest on funds held in bank accounts of community unless separate regulations provide otherwise; 11) grants from the budgets of other local government units; 12) other earnings community under separate regulations 2. The amount of the share in the proceedings of the tax on personal income from the tax payers living in the territory of the community is 39.34 %, subject to Art. 89 transitional provision 3. The amount of the share in the proceeds of the income tax law, which pay this tax,and located on the area of community, is 6.71 %	10) grants from the budgets of other local government institutions; 11) other county earnings under separate regulations. 2. the amount of the share in the proceeds of the tax on personal income from the tax payers living in the area of the county is 10.25 %. 3. the amount of the share in the proceeds of the income tax law, the taxpayers of this tax, both established in the area of the county, is 1.40 %	8) interest on funds held in bank accounts of the province, unless separate regulations decide otherwise; 9) grants from the budgets of other local government institutions; 10) other earnings of the province under separate regulations 2. the share of the proceeds from the income tax from individuals, from the taxpayers of this tax resident in the province is 1.60 % 3. the amount of the share in the proceeds of the income tax law, the taxpayers of this tax, both established in the province, is 14.75 %

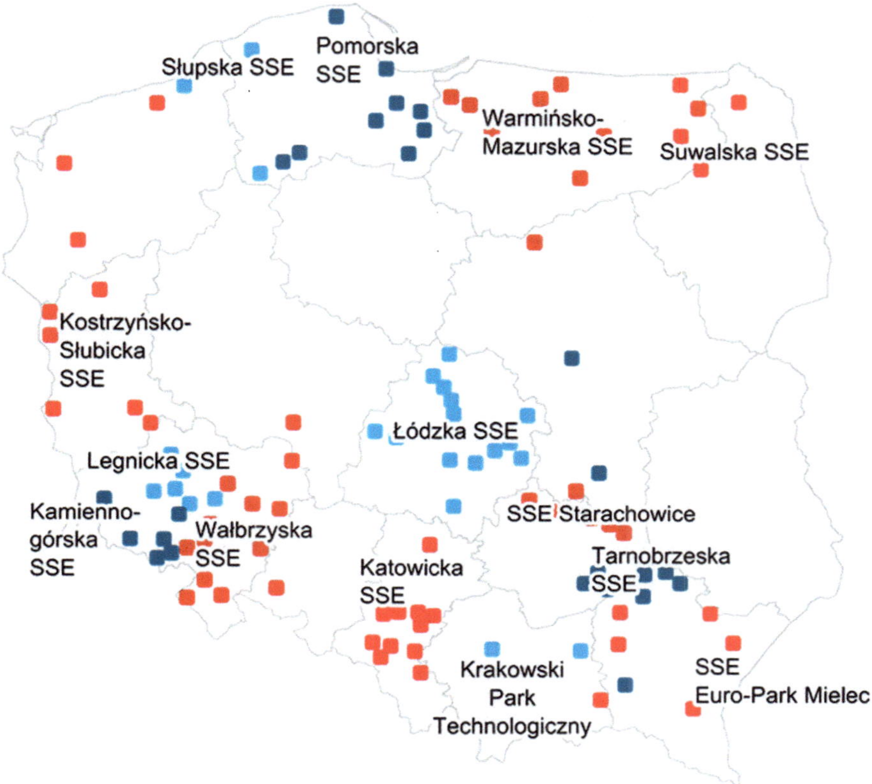

Fig. 2 Localization of special economic zones in Poland [8]

Other positive situations are if the local government decided to use a possibility which are in the new law acts, for example: the act of 8 September 2000. On commercialization, restructuring and privatization of "Polish State Railways" [10] and act of 28 March 2003. About railway transport [20].

The rule 9.3 of act from 29 March 2003 gives some regulations which enable the acquisition of railway lines by the local government [20].

3. Winding-up proceedings may be withheld if the competent local authority or designated by local government the entrepreneur:

1. Will provide funding to cover the costs uncovered by revenues from the provision from the rail carriers of this railway line or section of railway line;
2. Entered into an agreement for the acquisition of free railway line or section of railway line for the management for its continued operation;
3. Joined to the regional transport company which take over, in order to further exploit, provided to elimination the railway line or section of railway line, as a contribution in kind.

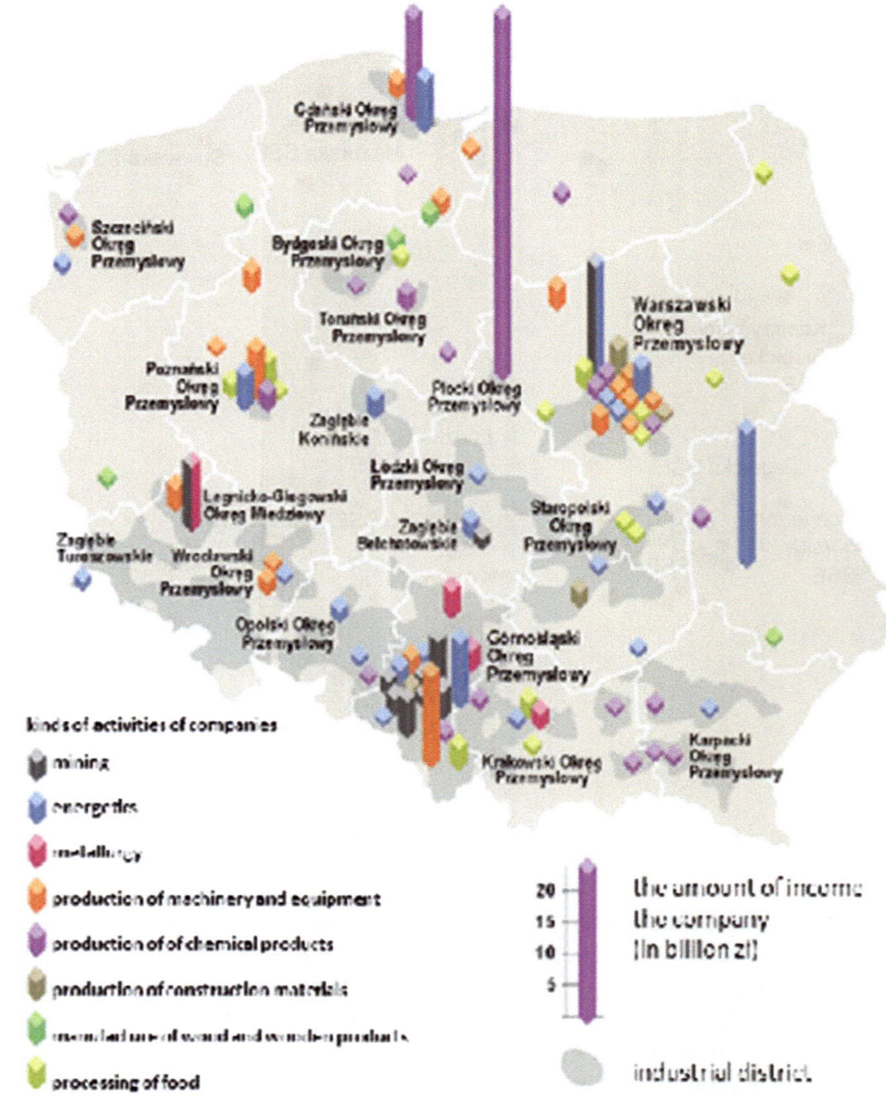

Fig. 3 Distribution of major industrial companies in Poland in 2008 [9]

Examples from Low Silesia

The acquisition of railway line 326 from Wrocław to Trzebnica in 2007 was the first step for sustainable transport strategy in Low Silesia. After reconstruction which lasted from 2007 to 2009, the first passenger train of local railway company Low Silesian Railways (pol.Koleje Dolnośląskie) could operate [11]. After 5 years of

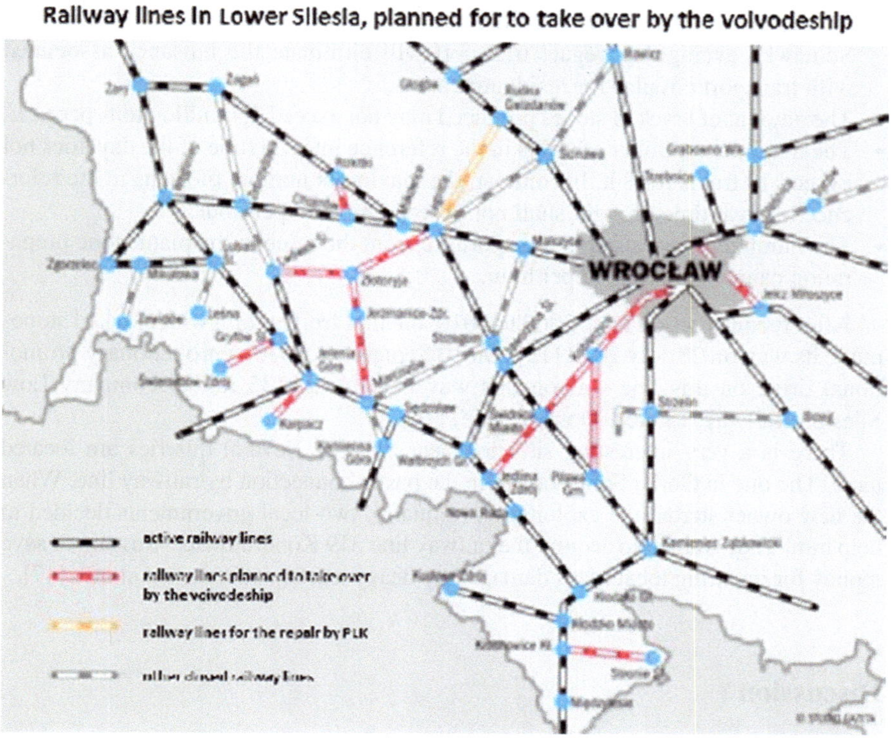

Fig. 4 Railway lines in Lower Silesia Studio Gazeta [13]

exploitation on this line, in 2014, it has been used by 240 thousand passengers each year to commute [12]. In 2014, Low Silesian parliament decided on acquisition of other railway lines (Fig. 4).

In the plans, is planned the regular passenger connection to cities and villages that would be attractive to tourists.

In the case of freight transport the first step was carried out by the local government of the small community Radków. This city is located in the Valley of Kłodzko ("Kotlina Kłodzka"). In 2006, the owner of a quarry of melaphyre applied to the local government of the municipality Radkow for environmental conditions to start deposits of stone exploitation. The municipality of Radków identified the environmental conditions [14]:

- It shall determine the transport of aggregates from the mine county road No. 3555D;
- Local needs in small amounts are allowed to transport from the mine provincial road that runs through the village Tłumaczów;
- Transport of the processed raw material outside the mine should be carried away in accordance with the conditions set out in the report on the impact of the proposed development of the field melaphyre "Tłumaczów–Gardzień field A"

on the environment, i.e., through the restoration of a railway Tłumaczów-Ścinawka average aggregate transport will eliminate the nuisance associated with transport circular for residential areas;

- The amount of cracked stones produced may not exceed 2.4 million tons per year;
- The maximum number of trains in the reference interval time of the day does not exceed 13 trains for 8 h. In contrast, the maximum number of trains in the reference interval time of night shall not exceed one train per hour;
- The number of cars used to transport ore from the mine to the plant stone preparation cannot extend 2.83 per hour.

After reconstruction from 2009 to 2010, the first freight train with cracked stones made its way on 28 July 2010 [15]. On 10 September, 2010, a probationary promotional drive on this line was on this way held by SA135 railcar company Low Silesian Railways (Koleje Dolnośląskie) [16].

There is a very interesting situation near Strzelin. Several quarries are located there. The one in Górka Sobocka had in the past a connection by railway line. When the new owner started the exploitation of quarry, two local governments decided to help him. They wanted to acquire the railway line 319 Kondratowice–Strzelin to save money for repairing local roads damaged by heavy trucks with cracked stones [17].

Discussion

The examples presented are based on the review of law acts and decisions. They showed that the present situation of the law does not close any possibility to decision for sustainable transport strategy in each level of local government.

In my opinion the future representatives of local government should be learned a methodology of the LCC (Life Cost Cycle) method. In this method all costs of technical system are analyzed. Present law act suggests to analyze only building cost. This is only part of all cost but no total cost of exploitation of the technical system.

Examples from a few local governments show that the calculation of budget of local government should begin not only for the time of building but also for time of exploitation.

This is a very good new stream of managing in local government.

There will be a better situation when the methodology of LCC is put into law acts.

Conclusion

Local government does not have to be a hostage in competition between different modes of transport. If representatives of local government decide to use modern tools for cost analysis, the total cost of exploitation of technical infrastructure in their territory will be lower.

The law acts and the training of local government representatives in analysis of total cost of exploitation of each technical infrastructure are the key factors for this.

Sustainable transport strategy is not area of interest only for the highest level of local government. Each level of local government may decide and should have possibility to do it. It depends only on the point of view of local representatives. If they do not see only one mode of transport, this local government can be presented as an ecological region.

References

1. Constitution.(1997): Constitution of the Republic of Poland, Dz.U. 1997 nr 78 poz. 483, Konstytucja Rzeczypospolitej Polskiej z dnia 2 kwietnia 1997 r. uchwalona przez Zgromadzenie Narodowe w dniu 2 kwietnia 1997 r., przyjęta przez Naród w referendum konstytucyjnym w dniu 25 maja 1997 r., podpisana przez Prezydenta Rzeczypospolitej Polskiej w dniu 16 lipca 1997 r.
2. Act of community.(1990): Dz.U. 1990 nr 16 poz. 95. Ustawa z dnia 8 marca 1990 r. o samorządzie terytorialnym.
3. Act of county.(1998): Dz.U. 1998 nr 91 poz. 578. Ustawa z dnia 5 czerwca 1998 r. o samorządzie powiatowym.
4. Act of province,(1998): Dz.U. 1998 nr 91 poz. 576. Ustawa z dnia 5 czerwca 1998 r. o samorządzie województwa.
5. lukasg: Krótka analiza polskiego rynku transportu drogowego, 18-07-2012, https://www.gieldafrachtowa.pl/krotka-analiza-polskiego-rynku-transportu-drogowego.
6. Stefaniak P.: Liberalizacja transportu pasażerskiego udała się tylko częściowo, 2012-07-12, http://www.obserwatorfinansowy.pl/forma/rotator/liberalizacja-transportu-pasazerskiego-udala-sie-tylko-czesciowo/.
7. Act about special economic zones pol. Ustawa z dnia 20 października 1994 r. o specjalnych strefach ekonomicznych.Dz.U. 1994 nr 123 poz. 600.
8. invest-in-wroclaw.pl: Specjalne Strefy Ekonomiczne, Lokalizacja głównych Specjalnych Stref Ekonomicznych w Polsce, http://www.invest-in-wroclaw.pl/doing-business/system-zachet-inwestycyjnych/specjalne-strefy-ekonomiczne/.
9. Wydawnictwa Edukacyjne WIKING 2005-2008: PRZEMYSŁ w Polsce, ROZMIESZCZENIE NAJWIĘKSZYCH FIRM PRZEMYSŁOWYCH w Polsce, http://www.wiking.edu.pl/article.php?id=288.
10. Act of 8 September 2000. On commercialization, restructuring and privatization of "Polish State Railways". (pol) USTAWA z dnia 8 września 2000 r. o komercjalizacji, restrukturyzacji i prywatyzacji przedsiębiorstwa państwowego "Polskie Koleje Państwowe". Dz.U.2000.84.948(Dz. U. z dnia 12 października 2000 r.).
11. Dolnośląska Służba Dróg i Kolei we Wrocławiu: Pierwszy przejazd szynobusu z Wrocławia do Trzebnicy. 2009-09-22, http://dsdik.wroc.pl/dsdik/aktualnosci/672-pierwszy-przejazd-szynobusu-z-wrocawia-do-trzebnicy.html.
12. Łapiński M.: Blisko ćwierć miliona pasażerów rocznie w szynobusach Kolei Dolnośląskich z Trzebnicy!, 23 września 2014, http://www.mareklapinski.pl/2014/09/koleje_dolnoslaskie_szynobus_trzebnica_wroclaw/.
13. Kokoszkiewicz M.: Pociągi wrócą na zamknięte linie kolejowe. Tory przejmie...2014-05-24, Gazeta.pl, http://m.wroclaw.gazeta.pl/wroclaw/1,121414,16022456,Pociagi_wroca_na_zamkniete_linie_kolejowe__Tory_przejmie___.html.
14. Decyzja Burmistrza Miasta i Gminy Radków w dniu 31 .05.2006 r. nr GPiOŚ-7624/13-4/06 w sprawie określenia środowiskowych uwarunkowań zgody na realizację przedsięwzięcia pod

nazwą: Eksploatacja złoża melafiru"Tłumaczów - Gardzień l" na działce nr 570/24 we wsi Tłumaczów gmina Radków".
15. Maciejewski M.: Nowy kamieniołom, 28 lipca 2010 r., tvp.pl.
16. Maciejewski M.: Próbny przejazd, 10 września 2010 r., tvp.pl.
17. Telewizja Strzelin: Gminy Strzelin i Kondratowice chcą przejąć i odtworzyć odcinek nieczynnej linii kolejowej nr 319, łączącej Strzelin z Kondratowicami. 14–05-2014, http://telewizjastrzelin.pl/2014/05/gminy-strzelin-i-kondratowice-chca-przejac-i-odtworzyc-odcinek-nieczynnej-linii-kolejowej-nr-319-laczacej-strzelin-z-kondratowicami/.
18. Kublik A.: KE nie da Polsce przesunąć na drogi pieniędzy, których kolej nie wyda?, 30.03.2012. http://wyborcza.biz/biznes/1,101562,11447717.html.
19. Stefaniak P.: Unia Europejska preferuje kolej, my tracimy miliardy niszcząc drogi, 2013-03-28 06:01, http://www.obserwatorfinansowy.pl/tematyka/makroekonomia/unia-europejska-preferuje-kolej-my-tracimy-miliardy-niszczac-drogi/.
20. Act of 28 March 2003. About railway transport. (pol) USTAWA z dnia 28 marca 2003 r. o transporcie kolejowym, Dz.U. 2007 nr 16 poz. 94, 2007-01-19.

EU Transport Policy Failure: The Case of Germany's Mindestlohngesetz

Wojciech Paprocki

Abstract EU transport policy represents an element of European economic strategy aimed at boosting growth in the single market. Among the goods and services that are traded in the market are international road transport services. The adoption of regulations on minimum wage (MiLoG law) by Germany's parliament forces transport companies headquartered outside Germany to pay no less than the German minimum wage to their drivers performing work on German territory. This legislation therefore breaches the EU principle of free movement goods and services and compels non-German transport undertakings to align their employees' pay to German rather than their native country's economic conditions. Arguably, the very fact that it was possible for a member state to introduce a regulation such as the MiLoG law should be seen as a EU transport policy failure.

Keywords EU transport policy • Minimum wage • Single market integration • International road transport

Introduction

Europe is busy looking for socioeconomic prescriptions to ensure that the benefits of continued economic growth are enjoyed by as many of the continent's inhabitants as possible. One of the preconditions is to strengthen the ability of Europe's economy to compete in the global marketplace. The European integration process, initiated in 1950 and originally involving only six Western European countries, has continuously evolved over the following decades, altering its organizational form and embracing new member states. The accession of Central and Southern European countries in 2004, 2007, and 2013, was of great consequence. The Union's eastward enlargement entailed new challenges, as the common market encompassed economies that were considerably less developed than

W. Paprocki (✉)
Department of Transport, Warsaw School of Economics, Warsaw, Poland
e-mail: wojciech.paprocki@sgh.waw.pl

M. Bąk (ed.), *Transport Development Challenges in the Twenty-First Century*, Springer Proceedings in Business and Economics,
DOI 10.1007/978-3-319-26848-4_6

those in the west of Europe, and the job market had to accommodate workforce accustomed to earnings that were much below the EU average.

The Community's economic policy comprises an array of sectoral policies, one of which addresses the transport sector and is conducted primarily via regulations targeted at the transport services market.

The introduction of a minimum wage law in Germany in 2015 (Mindestlohngesetz–MiLoG) signified a disturbance to the operation of the international transport market. Its impact was felt most intensely by transport undertakings headquartered in Central European countries, notably in Poland and Romania. This chapter discusses examples of such disturbances as well as their nature and repercussions.

The chapter aims to describe the circumstances in which the new regulation was adopted by Germany and to discuss the microeconomic and macroeconomic outcomes of its implementation. Given the number of publications that came out in the first quarter of 2015 delivering the findings of research focused on European law and national laws and examining the conformity of the German regulation with Community law, this article does not attempt to provide an extensive treatment of this aspect. In investigating the relevant issues, cause-and-effect analysis was utilized in combination with economic theory concerned with the design of employee compensation and motivation schemes. The author has also studied available statistics on unemployment rates and minimum wages, and looked at the factors influencing the purchasing power of employee incomes, including wages alongside other benefits.

Employment and Payment Across EU Economies

Sustainable development and job creation are among the key objectives of economic policy-making at the European level. These goals were literally spelled out in the Lisbon Strategy [1]. However, reality diverges significantly from what is articulated in political agendas, with registered unemployment among the population in productive age varying across regions, which is shown in Fig. 1.

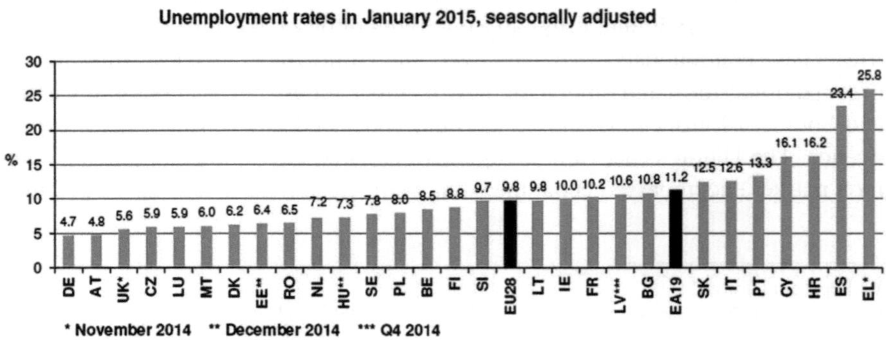

Fig. 1 Disparities in unemployment rate among EU member states (Source: Eurostat, Euro area unemployment rate at 11.2 %, March 2015)

It should be stressed that, even though unemployment rate statistics are generally indicative of demand for workforce in a particular job market, in some professions high unemployment rates (say, above 5.0 %) may actually concur with a shortage of workforce having certain qualifications (e.g., education and work experience). Such conditions can be traced back to an inappropriate structure of the vocational education sector that has been maintained for too long. For example, the current scarcity of professional drivers in Poland has been often attributed, among other factors, to the abolition of conscription; it is argued that in the second half of the twentieth century a large percentage of young men would be trained as drivers by the military. The expenses were at the time borne by the government, whereas now they have to be covered by the trainees themselves in the hope that they will be, at least partly, reimbursed by the prospective employer.

What deserves special attention is, alongside the availability of jobs, the question of minimum wages that are paid to the least skilled employees or in regions where there are a relatively small number of openings vis-à-vis the number of people in the productive age.

The issues of job availability and employee wages become more and more relevant all over Europe as the mobility of the population increases throughout the common market.

In Western Europe, e.g., in the Benelux countries, in France, Germany, the UK, and Ireland, the local workforce increasingly often have to compete for jobs with immigrants from new member states, mainly from Poland (the population of 38 million) and Romania (the population of 11.9 million) [2]. Objections to the inflow of workforce from these countries—which acceded in 2004 and 2007, respectively—were voiced even more forcefully after 2008, during the ongoing economic crisis, as relatively large groups of newcomers from both these countries (compared to the number of immigrants from other Central European countries) acquired jobs just because they were willing to accept lower wages than a local employee would.

Some Western European countries have already introduced the so-called minimum wage. This is the case of Germany where the relevant law came into effect on January 1, 2015, which was encouraged by political groupings advocating statist economy concepts based on the assumption that there are social and macroeconomic tradeoffs in regulating the labor market through for example laws mandating employers to pay the minimum wage.

The downsides of guaranteed minimum income have been repeatedly emphasized by the theory of economics [3] and demonstrated by economic practice. For more than 60 years, the government of the Federal Republic of Germany was unable to make the decision to introduce the minimum wage, not even under the rule of the social-democratic party (SPD). Instead, the issue was left to be decided by representatives of employers and trade unions under sectoral tariff agreements. In the 1950s and 1960s, Germany absorbed a wave of migrant workers, primarily from Spain and Italy, who received relatively low compensation for doing jobs that did not require highly specialist skills. The minimum wage was not brought into play at the time, nor was it afterwards, during the energy crisis of the 1970s and the following stagnation when no new jobs were created.

The only regulatory instrument that was then deployed was an administrative restriction on the number of job seekers arriving in Germany[1].

In the 2010s, the debate over minimum wage in Western Europe has been intertwined with a reflection on the growing income inequalities in the society that have emerged as a major economic challenge for the world's most developed countries as well as for many of the less developed ones. Governments seem to be committed to persuading business enterprises and their private owners to share their wealth. Economic poverty and its diverse effects are to be thus overcome [5]. Sociologists anticipate that we are about to witness one of the most spectacular clashes between public authorities and employers, engaging extremely powerful business organizations operating in the global market [6]. What is at stake is enforcing higher wages while avoiding the export of jobs outside Europe [7]. At the same time, governments seek to cause wages to rise throughout the economy more quickly than labor productivity. This is an intent that the theory of economics commonly sees as undermining the economic equilibrium and the sustainability of macroeconomic growth [8].

Almost as soon as the minimum wage was introduced in Germany, academics and business communities in EU member states became aware of something that had escaped their attention before 2014. Namely, it dawned upon some of them that the new law affected not only persons employed by business enterprises based in Germany and undertaking their jobs within the country, or individuals employed by business enterprises based outside Germany and sending their employees over to work in Germany, but also persons employed by foreign businesses who perform their duties while traveling across Germany. Article 16 of the MiLoG act stipulates that employers headquartered outside Germany are obliged to report (to the German government) any employees staying in the country [9]. Under this law, anyone, including for example professional drivers, who stays in the area of Germany even for a short time (e.g., in performance of a freight shipment from outside of Germany to a destination near Germany's border, provided that more than 2 h have elapsed since their crossing the border at a point of entry), must be compensated for their work at an hourly rate that is equal or higher than the statutory minimum rate of EUR 8.50.

Economic Disparities Among EU Member States: The Minimum Wage Aspect

As a result of the most recent enlargement in 2013, the European Union now comprises 28 member states. With the increasing number of member states, the common market has widened, too, which is seen as a factor conducive to economic growth throughout the Community.

EU member states are characterized by significant regional disparities. One way to capture these regional differences is through the minimum wage defined

[1] The "no entry" rule was put in place (*Anwerbestopp* 1973) [4].

by specific countries. Minimum wage differentials are analyzed by looking at two principal measures:

- the minimum wage,
- the purchasing power standard (PPS)[2].

The fundamental advantage of using the absolute value of minimum wage as a measure is that the disparities that are revealed are "unquestionable," having been prescribed in the national laws of specific member states. These are either a minimum monthly pay (e.g., in Poland—at PLN 1750 since January 1, 2015) or an hourly rate (e.g., in Germany—at EUR 8.50 since January 1, 2015).

To be able to make meaningful references to values in currencies used in specific member states, and to perform comparative analyses between them, researchers need to apply currency exchange rates. Observations of the volatility of currencies made over years as well as several-month periods indicate that the rates of exchange between EU member states' currencies tend to fluctuate dynamically and bidirectionally—these findings are illustrated in Fig. 2. Therefore, any estimates of differences in minimum wages should be frequently updated.

In their countries of residence, people who have the same financial resources in nominal terms are able to purchase goods (commodities or services, including

Fig. 2 Change dynamics of the EUR to PLN exchange rate (Source: own estimation based on the data published by the National Bank of Poland)

[2] PPS are defined by equating the total real final expenditure of the EU 25 on a specific basic heading, aggregate or analytical category to the total nominal final expenditure of the EU 25 on the same basic heading, aggregate or analytical category. Cf. [10].

public services subsidized by national or local governments) of diverse value. This phenomenon stems from the two key factors:

- The same goods may be priced differently in the markets of specific member states, and in some market segments the differences may be even dramatic (e.g., municipal public transport services in Warszawa, Poland and in Frankfurt am Main, Germany—in March 2015, a single ride pass cost PLN 4.40 [11] and EUR 2.75 [12], respectively, while the EUR/PLN exchange rate is 4.1255 [13], which means that a corresponding zone-based fare was 158 % higher in the German city than in the Polish capital,
- In their place of residence, people are familiar with the characteristics of the local market and are capable of purchasing a number of goods, including foods, from their preferred outlets where most or all goods are priced below average, and at the same time avoid buying from shops where the prices are above average, whereas for example foreign travelers are not always in the position to explore the market well enough, due to the language barrier and other obstacles, to make their own price comparisons and refrain from making highly priced purchases.

Differences in the nominal value of minimal wage across some EU member states are shown in Fig. 3, while the disparities in purchasing power standard (PPS) are shown in Fig. 4. Figure 5, on the other hand, by bringing together different measures of economic differentials across member states, makes it possible to capture substantial discrepancies between estimates produced by particular tools.

In lieu of this discussion and the examples provided, it can be concluded that in assessing the competitiveness of specific member states (the macroeconomic perspective) or the competitiveness of business enterprises headquartered in specific countries (the microeconomic perspective) one should not rely on a single indicator involving the use of minimum wage differentials only.

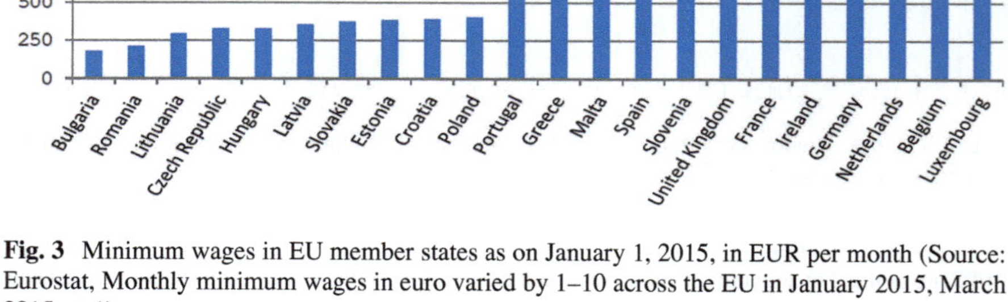

Fig. 3 Minimum wages in EU member states as on January 1, 2015, in EUR per month (Source: Eurostat, Monthly minimum wages in euro varied by 1–10 across the EU in January 2015, March 2015, p. 1)

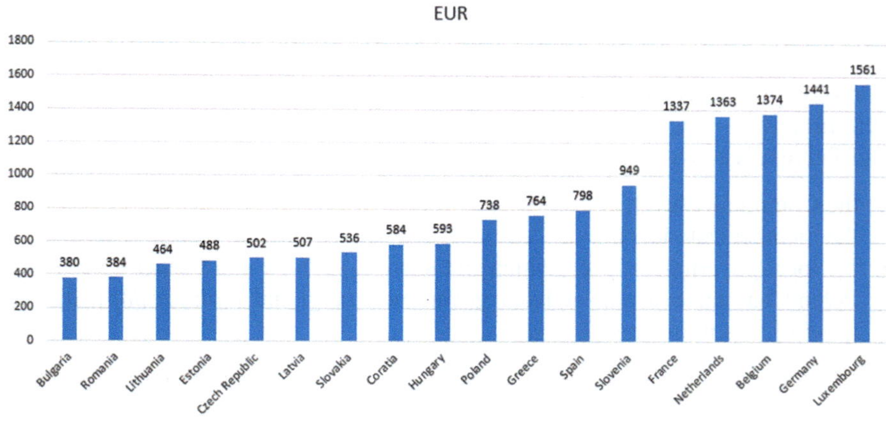

Fig. 4 Minimum wages per month in purchasing power standards in EU member states as on January 1, 2015 (Source: own based on Eurostat, Monthly minimum wages in euro varied by 1–10 across the EU in January 2015, March 2015, p. 3)

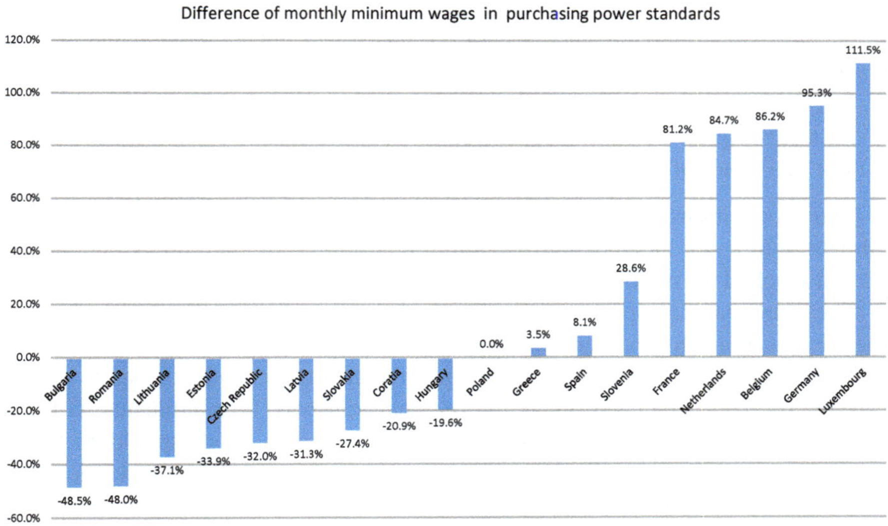

Fig. 5 Differences in monthly minimum wages in purchasing power standards in the EU member states as on January 1, 2015 (Source: own)

EU Transport Policy

Transport policy epitomizes sectoral policies pursued by what could be described as "the Empire"[3]—a term that designates the European Union's specific system of legislative, executive and judicial power. It is composed of three levels:

- the Community level,
- the member state level,
- the local government level (which can itself be divided into further levels, e.g., Poland has three levels of local government: regional [*voivodeship*], district [*poviat*], and municipal [*gmina*]).

EU transport policy is delineated in documents that are referred to as White Papers. The 2011 White Paper outlines a transport agenda identifying the key measures to be taken and putting forth the vision of a competitive and sustainable transport system that is made up of the following components:

- Growing Transport and supporting mobility while reaching the 60 % emission reduction target,
- the efficient core network for multimodal intercity travel and transport,
- the global level-playing field for long-distance travel and intercontinental freight,
- clean urban transport and commuting,
- Ten Goals for a competitive and resource efficient transport system: benchmarks for achieving the 60 % GHG emission reduction target.
- The Ten Goals of EU transport policy are detailed in Table 1.

The propositions of the 2011 White Paper are in line with the goals of EU climate policy, the latter being essentially determined by the 2009 Copenhagen accord reached during the international talks around the Kyoto protocol [20]

The European transport system is founded on preference for rail, which is regarded as more environment friendly than road transport [21]. The policy toward road transport is grossly inconsistent. In general, efforts are made to confine freight carriage by road vehicles to local and regional transport, i.e., shipments within a radius of about 300 km. Over distances longer than this, goods should be transported by rail, and where rail operators cannot offer affordable direct services below the minimum weight required for block trains, freight should be carried via combined rail-road transport [22]. The latter technology can reduce the use of cars to regional movement and delivery of goods in unit loads (e.g., containers), by trailers or swap-bodies, from the sender's premises to a terminal or from a terminal to the recipient's premises.

At the same time, EU member states, including Poland, proceed with road infrastructure development projects, and the automotive industry, even now ranked among the strongest sectors of Europe's economy, benefits from political and financial support (e.g., tax exemptions) that allows it to further boost its potential. Both

[3] For a broader discussion of the functions of the Empire, see [14].

Table 1 The Ten Goals envisaged by the 2011 European Commission White Paper

Developing and deploying new and sustainable fuels and propulsion systems
1. Halve the use of "conventionally-fuelled" cars in urban transport by 2030; phase them out in cities by 2050; achieve essentially CO_2-free city logistics in major urban centres by 2030[a]
2. Low-carbon sustainable fuels in aviation to reach 40 % by 2050; also by 2050 reduce EU CO_2 emissions from maritime bunker fuels by 40 % (if feasible 50 %)[b]
Optimising the performance of multimodal logistic chains, including by making greater use of more energy-efficient modes
3. 30 % of road freight over 300 km should shift to other modes such as rail or waterborne transport by 2030, and more than 50 % by 2050, facilitated by efficient and green freight corridors. To meet this goal will also require appropriate infrastructure to be developed
4. By 2050, complete a European high-speed rail network. Triple the length of the existing high-speed rail network by 2030 and maintain a dense railway network in all Member States. By 2050 the majority of medium-distance passenger transport should go by rail
5. A fully functional and EU-wide multimodal TEN-T "core network" by 2030, with a high quality and capacity network by 2050 and a corresponding set of information services
6. By 2050, connect all core network airports to the rail network, preferably high-speed; ensure that all core seaports are sufficiently connected to the rail freight and, where possible, inland waterway system
Increasing the efficiency of transport and of infrastructure use with information systems and market-based incentives
7. Deployment of the modernised air traffic management infrastructure (SESAR[c]) in Europe by 2020 and completion of the European Common Aviation Area. Deployment of equivalent land and waterborne transport management systems (ERTMS[d], ITS[e], SSN and LRIT[f], RIS[g]). Deployment of the European Global Navigation Satellite System (Galileo)
8. By 2020, establish the framework for a European multimodal transport information, management and payment system
9. By 2050, move close to zero fatalities in road transport. In line with this goal, the EU aims at halving road casualties by 2020. Make sure that the EU is a world leader in safety and security of transport in all modes of transport
10. Move towards full application of "user pays" and "polluter pays" principles and private sector engagement to eliminate distortions, including harmful subsidies, generate revenues and ensure financing for future transport investments

Source: WHITE PAPER: Roadmap to a Single European Transport Area – Towards a competitive and resource efficient transport system, COM (2011) 141 final

[a]This would also substantially reduce other harmful emissions

[b]Cf. Communication from the Commission to the European Parliament, the Council, the European Economic and Social Committee and the Committee of the Regions "A Roadmap for moving to a competitive low carbon economy in 2050," COM (2011)112[15]

[c]In accordance with the European ATM Master Plan [16]

[d]In accordance with the European Deployment Plan for ERTMS: cf. Commission Decision C(2009)561[17]

[e]In accordance with the EasyWay 2 Implementation Plan: cf. Commission Decision C(2010) 9675

[f]Directive 2002/59/EC establishing a Community vessel traffic monitoring and information system (OJ L 208 of 5.8.2002), as amended by Directive 2009/17/EC (OJ L 131 of 28.5.2009) [18]

[g]Cf. Directive 2005/44/EC [19]

EU institutions and national governments are well aware of the ongoing growth of e-commerce and try to respond to it by addressing its immediate effects, such as for example the soaring demand for courier services—that are mostly performed using light commercial vehicles.

A salient characteristic of the European transport system is the presence of multiple regulations constraining the behavior of service providers and hence limiting the freedom of forwarders (i.e., entities placing freight shipping orders) to choose among carriers [23].

The rationale for introducing market regulations is typically either or both of the two:

- to influence the market by creating barriers to entry or imposing constraints on certain operations and behaviors in an attempt to shift traffic between modes of transport toward the modal split that is preferred by the regulator;
- to prevent or discourage service providers from certain behaviors that might incur excessive risk, e.g., cause heavier traffic that increases the risk of vehicle-to-vehicle accidents as well as of collisions between vehicles and persons or objects in their closest vicinity.

In the 1980s, some European Alpine countries, viz. Austria and Switzerland, decided that it was socially, economically and environmentally desirable to reduce road traffic. Facing a substantial growth in trade volume, they sought to carry as much freight as possible by rail. In Switzerland, a symbolic ceiling of one million vehicles annually was established for road transit. This objective, albeit backed up with a variety of restrictions on road transport and preferential mechanisms for rail transport, was not achieved, as demand for road transit continued to rise steadily until the traffic exceeded the threshold. The authorities had to admit that the policy was ineffective. It was realized that administrative constraints on the growth of road transport threatened to arrest the country's economic growth by damping interregional trade.

Regulations concerning the behavior of transport undertakings and their personnel, primarily drivers, aim to enhance road safety by minimizing the risk of road accidents attributable to fatigue. Standards applicable to drivers' working conditions are termed in EU law as social rules in road transport or social legislation relating to road transport. Under the pretense of social legislation, specific EU member states have put in place—on top of EU regulations—even more stringent national laws whose actual purpose is to protect national job markets against the expansion of transport companies employing drivers from Central European countries, including Poland [24]. Such moves are not consistent with the EU policy set out in the White Paper and clearly demonstrate the increasing prevalence of national priorities over Community interests [25]. Examples of such measures include the legislation that came into force on July 1, 2014 in Belgium and France, and requires professional drivers to spend the so-called long weekend breaks at a hotel [26]. The regulation is supposed to ensure better resting conditions for drivers than the usual practice of sleeping and relaxing in a truck or tractor cabin.

The ability of Polish companies to compete in the international road transport services market has been conditioned by their modest operating costs. These were minimized by keeping driver remuneration at the national minimum wage level, on top of which were paid, at least partly, the mandatory travel allowances. At the moment, there are many formal disputes regarding the payment of travel allowances that are to be settled in court. If the disputes are resolved in the employees' favor, some transport undertakings may have to bear additional costs in the years to come. And, if that is the case, the competitive edge of Polish transport companies may well be lost.

There are a number of factors relating to the pursuit of EU transport policy that might adversely affect the competitive position of Polish road transport companies in the common market. Prior to 2014, it was hardly possible to predict the current trend of changes [27]. Two unexpected developments took place in 2014:

- contrary to long-term forecasts, the prices of liquid fuels plunged globally, bringing down the price of diesel fuel, too;
- the parliament of Germany passed the MiLoG law.

The dramatic decline of liquid fuel prices disturbed the equilibrium pricing process in the international market for freight road transport services. Isolated stand-alone companies that represent the supply side of the market are at a disadvantaged position in bargaining with huge industrial and commercial conglomerates constituting the demand side, therefore transport undertakings fear strong pressures on lowering freight rates, which could further reduce their profit margins.

The implementation of the MiLoG regulation may have a variety of implications, some of which have not at all been investigated by its proponents. This means that very little is known about the likely effects of the German minimum rates in case they come to be applied widely by international businesses.

MiLoG: An Instrument Alien to EU Transport Policy

The socioeconomic growth in the EU area depends on the pace of market integration. The more effectively specific nations and professional groups can safeguard their individual interests in national markets, the fewer beneficiaries of the single market there will be across the European Union. It will not suffice then to promote the Union through ads and slogans; it is necessary for EU bodies to deploy more broadly solutions supporting effective market integration and to observe strict discipline in implementing EU policies [28]. The adoption of the MiLoG law by the German parliament should be seen as representing an interference to the EU market integration process, since it steps into the area of regulation that should be agreed on and implemented solely at the EU level. Transitional periods that were defined on EU enlargements in 2004 and in the following years enabled both old and new member states to adjust to new standards [29]. In many areas, however, including the job market, the adaptation process has not finished yet [30]. Substantial problems arise in Western European countries where the second decade of the twenty-first century has been marked with increasingly frequent attempts to launch, or extend the use of, protections for the domestic market. This is inevitably going to reverse what has been achieved with regard to the freedom of trading in the EU market—both in goods and in services.

An important implication of the regulations laid down by the MiLoG law is Germany's parliament direct intervention into national legislation enacted in other EU member states. It not only disrupts business management processes in undertakings headquartered outside Germany but also impinges on the powers of national authorities in all other (except Germany) countries of the EU. The MiLoG law had been in effect for just a few weeks when the first complaints were made that the obligation to report each entry into Germany, imposed on all foreigners who, being in employment with companies registered outside Germany, performed work temporarily in Germany, involved an immense administrative burden [31, p. 2]. It is also pointed out that penalties for infringements of the law are exorbitant and incommensurate with the harm that such infringements might cause ([31], p. 3). Furthermore, objections are raised as to empowering the German government to control and influence the activities of business enterprises domiciled outside Germany through legislation that was enacted unilaterally by German parliament.

Harmonization of social conditions for all employees throughout the EU involves eliminating distortions to fair competition between businesses registered in different EU members states that result from striking differences in working conditions and disproportions in wages. Regulations of the MiLoG law were designed to function within German economy and may be incompatible with solutions that are found in other countries, such as for example Poland, which continues to abide by a number of regulations originating in the period of administrative command economy. What the MiLoG law ignores is for example the fact that Polish (as well as several other member states') compensation systems and Polish tax law place limits on employee benefits by specifying threshold amounts for payments that can be made toward a non-wage component. The trickiest part is, however, the regulations on the payment of benefits in foreign currencies and their treatment under personal income tax schedules available to individual employees. Furthermore, a significant part of professional driver's earnings is not a wage under the German MiLoG law but instead has a motivational function [32], since it is presumed to have a certain stimulating effect on the employee. The German lawmakers disregarded the economic relevance, for transport companies as well as for drivers, of that part of the employer's costs which is considered an element of wages in Germany but is not elsewhere. A portion of the compensation paid to drivers for example in Poland is not included in the calculation of the actual pay and, consequently, is not subject to compliance checks against minimum wage legislation.

Moreover, in the context of minimum wages, it does not seem economically sound to compare earnings received by employees residing outside Germany and staying temporarily in the country in performance of their professional duties with those of employees having their permanent residence and employment in Germany in terms of nominal value rather than purchasing power. This is certainly a structural flaw inherent in the regulations laid down by the MiLoG law.

The pressure on increasing the nominal wages of drivers employed by transport undertakings headquartered outside Germany has implications for the transport industry throughout the single market, affecting not only the international transport market, but national transport markets as well. Since professional drivers can choose between employment with freight and passenger transport businesses, including urban public transport undertakings, increasing the wages of drivers in the international transport sector may create pressures on raising nominal wages in other businesses, too. What raises most controversies is the pressure on pay rises in public transport companies, whose operating expenses, including payroll, are not wholly financed from fares paid by individual customers. In Poland and other Central European countries, as well as in Western Europe, public companies are heavily reliant for their budgets on government subsidies (mostly from local governments that are in turn backed up by central governments and may even be cofinanced from EU funds). It seems that, in shaping the provisions of the MiLoG law, the parliament of Germany did not give enough consideration to the ways in which the regulation could bear on, e.g., operating costs in the entire transport sector, including subsidized public transport companies.

When taking into considering the above, it can be concluded that the regulations implemented through the MiLoG law are defective and alien to the European system. Hence, there are grounds to believe that their enforcement will inhibit rather than assist the EU integration process or the achievement of EU transport policy objectives.

Conclusions

In drafting the act on minimum wage (MiLoG law), the legislature sought to reduce income inequalities by regulating the lowest wages throughout the Federal Republic of Germany. Its provisions apply to everyone who performs work on German territory, including drivers employed by business enterprises based in other EU member states. It was not until the MiLoG came into effect on January 1, 2015 that it was perceived how the law fitted into the entrenched conflict of economic interests, dating back to 2004 and escalating over the years, between providers of national and international road transport services that are headquartered in Germany and those that are registered in other countries of the EU. The central question is whether the instrument, i.e., the MiLoG law, can be ultimately beneficial to the EU economy as a whole, or whether it is only going to reinforce the protectionist policies of German federal government that are obviously incompatible with the objectives of EU economic policy and transport policy.

The confusion surrounding the European Union's road transport services market reveals that, for a number of years, EU institutions have not been giving enough attention to the market's real concerns. Overwhelmed by different aspects of climate change, EU transport policy fails to offer anything but policy statements that are not subsequently translated into regulations facilitating growth on the supply as well as the demand side of the transport market.

Regretfully, deliberations on EU transport policy have been diverted away from the need for major adjustments toward the compensation of drivers in the international road transport sector. As a result, the MiLoG law, which is de jure applicable in a single country and which addresses issues that are beyond the scope of EU transport policy, forces European bodies to take a defensive stance, temporarily undercutting their role as part of the Empire in implementing initiatives to support economic growth across the EU. Cases where Community-wide regulations are transposed into national legislation in order to, somewhat paradoxically, satisfy the needs of specific member countries should be assessed as manifestations of harmful practices that are brought within the framework of European integration policy and EU transport policy.

References

1. Community Research and Development Information Service, http://cordis.europa.eu/programme/rcn/843_en.html (retrieved on March 21, 2015).
2. Eurostat, http://ec.europa.eu/eurostat/tgm/table.do?tab=table&language=en&pcode=tps00001&tableSelection=1&footnotes=yes&labeling=labels&plugin=1 (retrieved on March 21, 2015).
3. Nell, G. L. (ed.), Basic Income and the Free Market. Austrian Economics and the Potential for Efficient Redistribution. Palgrave Macmillan, New York (2013), p. 3 and onwards.
4. Eurostat, http://ec.europa.eu/eurostat/tgm/table.do?tab=table&language=en&pcode=tps00001&tableSelection=1&footnotes=yes&labeling=labels&plugin=1 (retrieved on March 21, 2015).
5. Broda-Wysocki, P., Wykluczenie i inkluzja społeczna. Paradygmaty i próby definicji, p. 102 and following. Instytut Pracy i Spraw Socjalnych, Warszawa (2012).
6. Kunze, A., Lebert, S., Malter, B., Zimmermann, F., Mindestlohn. Lohnlüge. Die Zeit 10 (2015).
7. Radło, M.-J., Offshoring i Outsourcing. Implikacje dla gospodarki i przedsiębiorstw, p. 178. Oficyna Wydawnicza Szkoła Główna Handlowa w Warszawie, Warszawa (2013).
8. Gruchelski, M., Bezrobocie w procesie wzrostu gospodarczego. Teoria i praktyka, p. 39. Oficyna Wydawnicza Szkoła Główna Handlowa w Warszawie, Warszawa (2013).
9. Gesetz zur Regelung eines allgemeinen Mindestlohns (Mindestlohngesetz – MiLoG) vom 11. August 2014, BGBl. I S 1348.
10. OECD, http://stats.oecd.org/glossary/detail.asp?ID=7184 (retrieved on March 21, 2015).
11. Zarząd Transportu Miejskiego w Warszawie, http://www.ztm.waw.pl/?c=110 (retrived on March 21, 2015).
12. Rhein-Main-Verkehrsverbund, http://www.rmv.de/auskunft/bin/jp/query.exe/dn?ld=14.103&seqnr=1&ident=hi.029225103.1426938120 &CMS_AppId=Tarifauskunft&OK#focus (retrieved on March 21, 2015).
13. Narodowy Bank Polski (National Bank of Poland), http://www.nbp.pl/ (retrieved on March 21, 2015).
14. Mazur, B., Paprocki, W., Rola Imperium w kształtowaniu i realizacji polityki gospodarczej, ze szczególnym uwzględnieniem polityki energetycznej. In: Gajewski, J., Paprocki. W. (eds), Dylematy rozwoju infrastruktury, , p. 13-onwards. Proceedings of the European Financial Congress, Gdańsk (2014).
15. Communication from the Commission to the European Parliament, the Council, the European Economic and Social Committee and the Committee of the Regions "A Roadmap for moving to a competitive low carbon economy in 2050," COM (2011)112.

16. European Commission, http://ec.europa.eu/transport/air/sesar/deployment_en.htm (retrieved on March 21, 2015).
17. Commission Decision C(2009)561.
18. Directive 2002/59/EC establishing a Community vessel traffic monitoring and information system (OJ L 208 of 5.8.2002), as amended by Directive 2009/17/EC (OJ L 131 of 28.5.2009).
19. Directive 2005/44/EC.
20. Gregori, G., Wimmer Th. (eds.), Grünbuch der nachhaltigen Logistik. Handbuch für die ressourcenschonende Gestaltung logistischer Prozesse, p. 19. Bundesvereinigung Logistik, Wien-Bremen (2011).
21. Mężyk, A., Uwarunkowania i efekty reform kolei, , p. 157. Politechnika Radomska, Radom (2011).
22. Wronka, J., Transport kombinowany/intermodalny. Teoria i praktyka, p. 33. Wydawnictwo Naukowe Uniwersytetu Szczecińskiego, Szczecin (2009).
23. Załoga, E., Trendy w transporcie lądowym Unii Europejskiej, p. 60. Wydawnictwo Naukowe Uniwersytetu Szczecińskiego, Szczecin (2013).
24. Aberle, G., Transportwirtschaft. Einzelwirtschaftliche und gesamtwirtschaftliche Grundlagen, 4th ed., p. 190. Oldenbourg, München, Wien (2003).
25. Meyer, M., Alten und neue Fronten in Europa. Rückkehr zur Nationalismus? Neue Zürcher Zeitung (2014), http://www.nzz.ch (retrieved on April 27, 2014).
26. Rynasiewicz, Z., Comments on the road haulage market debate. Europolitics 4893, 16 (2014).
27. Rydzkowski, W., Matusiewicz, M., Szanse i zagrożenia międzynarodowego transportu drogowego w Polsce. In: Funkcjonowanie i rozwój transportu. Zeszyty Naukowe Ekonomika Transportu Lądowego 41, 90. Wydawnictwo Uniwersytetu Gdańskiego, Gdańsk (2011).
28. Lilla, M., Eine Weltordnung ohne Europa. Neue Zürcher Zeitung 2015, http://www.nzz.ch (retrived on March 23, 2015).
29. Barcz, J. (ed.), Ustrój Unii Europejskiej, p. VII-62-63. Instytut Wydawniczy EuroPrawo, Warszawa (2010).
30. Böning, M., Europäische Arbeitsrechtsangleichung zwischen Anspruch und Wirklichkeit, p. 223. Nomos, Baden-Baden (2011).
31. DSLV-Positionspapier: Dringende Korrektur des Mindestlohngesetzes (MiLoG) erforderlich. DLSV, Berlin (2015).
32. Nieckarz, Z., Psychologia motywacji w organizacji, pp. 76-77. Difin, Warszawa (2011).

Costs and Benefits of High Speed Rail Integration in Europe

Anna Urbanek

Abstract Undertakings aiming at integration in the high speed rail sector require capital-intensive and time-consuming investments in infrastructure, modern rail traffic managements systems, or ICT technologies supporting the process of booking and selling tickets. It also requires covering the expenses related to the process of preparation and management of contracts and agreements. This chapter attempts to identify and quantify the costs and benefits of integration projects undertaken in the field of high speed passenger transport in Europe. Cost–benefit analysis, including external cost and benefits, is essential for assessing the effectiveness of investment and it is especially important in the case of investments financed with the use of public funds.

Keywords High speed rail • Strategic alliance • ICT • Interoperability • Transport integration • ECTS • ERTMS • Railteam

Introduction

In Europe it is assumed that the borderline speed, separating high speed rail (HSR) from traditional one, is the speed of 250 km/h, which is possible to be achieved on lines built especially to be used by high speed trains, or lines suitably modernized for that purpose [1]. The construction of HSR requires not only investment in suitable infrastructure and rolling stock, but also in Information and Communications Technologies (ICT), such as modern rail transport management systems, or technologies supporting dynamic passenger information or the process of booking and selling tickets.

In the European Union high speed rail transport is the most dynamically developing passenger transport segment, which is very important, especially in the context of strong decline of traditional rail transport demand for over 20 years. Thereby, high speed rail transport is seen as a chance to revitalize passenger rail transport in Europe.

A. Urbanek (✉)
Department of Transport, University of Economics, Katowice, Poland
e-mail: anna.urbanek@ue.katowice.pl

M. Bąk (ed.), *Transport Development Challenges in the Twenty-First Century*,
Springer Proceedings in Business and Economics,
DOI 10.1007/978-3-319-26848-4_7

European Union policy supports the integration of national high speed rail systems into trans-European network. It is one of the means used to achieve the most important aims of the Community. The process of integration of high speed rails in Europe will absorb an important volume of public funds in the following years. According to the assessment of the European Commission, the entire system of high speed rail, both the newly constructed lines that are suitable for the speed of 250 km/h and auxiliary modernized lines, should by 2020 reach the length of 23,198 km (as early as in 2017—more than 20,000 Km), thus roughly twice as much as at present [2].

Every time an investment decision has to be taken, one or another form of weighing costs and benefits is involved. This applies in particular to the investments financed using public funds, including investment projects applying from the EU Funds [3]. They require identification and assessment of costs and benefits of the undertaking in a definite time horizon, taking into account the external costs and benefits, which are paid or gained by the surroundings, including the natural environment and society. The aim of the chapter is to try to identify the costs and benefits of integrating high speed rails in Europe.

Identification of Costs Related to Integration of Entities Providing High Speed Rail Services in Europe

Integration of passenger transport is most often defined as an organization process, by which the elements of passenger transport system (network and infrastructure, fares and tickets, information and marketing, etc.) of various branches of transport and service providers cooperate ever more closely and more efficiently, the result of which is the increasing quality of public transport services, coupled with elements of individual travels [4].

Among the most frequently encountered costs of transport integration, which burden the supply side, the literature of the subject most frequently lists the following ones [5]:

- costs related to planning an integrated transport network,
- costs of preparation and maintenance of contracts and agreements between entities involved in the integration process, and public entities supervising a given market,
- cost of IT systems,
- costs of traffic management in unexpected and emergency situations,
- costs of maintenance and management of personnel providing services to customers, in normal traffic conditions and in case of disturbances/emergencies,
- costs related to standardization of information for passengers and standards of services provided.

Integration of high speed rails in Europe is a process, which takes place in two dimensions, namely the technical dimension, which consists of harmonization of technical infrastructure, and the organizational dimension, e.g. concerning the fare and ticketing integration, or co-ordination of timetables. Technical harmonization of infrastructure is related firstly to striving towards interoperability of rail network in Europe, thus technical and operational unification. In that process, besides construction of proper infrastructure, of key importance is the development and implementation of a uniform—in the member states—system of rail traffic management and control—ERTMS (European Railway Traffic Management System). ERTMS is an indispensable element of HSR system, as it allows to control rail traffic at the speeds exceeding 200 km/h. The system is composed of two main subsystems: European Train Control System (ETCS) and GSM for Railways (GSM-R). Their functioning is possible thanks to devices, which enable transmission of information from the equipment installed along railway lines to the driver's cockpit, as well as provide voice communication e.g. with train dispatchers, as well as digital data transmission [6]. It is calculated that the construction of 1 km of HSR lines entails the cost of 12–30 million EUR (Table 1). The level of that cost depends first of all on the relief, population density, and the route of the line.

The yearly maintenance costs of HSR lines amount to some 70,000 EUR/km, which entails, for the operations of a high speed line having the length of 300 km, the yearly maintenance cost of about EUR 21 million. Additionally, one should also take into account the costs of purchase and maintenance of suitable rolling stock, as well as costs of construction and equipping the transfer modes. Also the costs of implementing the ERTMS system are substantial. Table 2 provides the costs of implementation of ETCS system, resulting from UIC estimates, as well as costs of implementation of ETCS observed in business practice, as a result of comparative studies conducted by UIC in various countries. The system ETCS may be implemented at two levels, which differ in terms of functions they have and technologies used. Figures in Table 2 suggest that the cost of implementing the ETCS system level 2 per 1 km of double track amounts to about EUR 500 thousand, while as the business practice demonstrates, in some cases it can be even twice as much.

The construction of a trans-European system of high speed rail comprises also the sphere of organization, e.g. via co-operation of rail companies as concerns co-ordination of timetables, co-operation regarding common tickets, services and information for passengers, or joint promotion of services. Those activities require the support of modern ICT technologies, thus the technical requirements for interoperability of the trans-European HSR system comprise also telematic applications (TAP-TSI), including—among other things—systems for providing information to passengers before and during the journey, systems for booking seats and making

Table 1 Magnitude of costs of high speed rail system [7]

No.	Cost category	Value (in EUR)
1.	Construction of 1 km of new HSR line	12–30 million
2.	Yearly maintenance costs per 1 km of HSR lines	70,000
3.	Purchase of a high speed train (350 places)	20–25 million
4.	Yearly maintenance costs of a high speed train (assuming 2 EUR/km and yearly operation work of 500,000 km/train)	1 million

Table 2 Costs of implementation of ETCS (level 1 and level 2) [8]

No.	Cost category	Level 1 (in EUR per 1 km of double track)	Level 2 (in EUR per 1 km of double track)	Installation of equipment on the train
1.	Studies, technical documentation, design, etc.	20,000	100,000	110,000
2.	Costs of infrastructure construction	90,000	400,000	340,000
3.	Costs of system implementation (total)	110,000	500,000	450,000
4.	Costs of system implementation observed in business practice	35,000–140,000	140,000–940,000	120,000–800,000
5.	Yearly costs of the system operation and maintenance	15,000	130,000	110,000

payments for services, or sales of tickets in ticket offices, automatic vending machines, and via mobile technologies [9].

One of the main challenges as regards integration of national HSR systems in the next few years is also the implementation of an integrated European booking and ticketing system. In the years 2007–2009, in the framework of the strategic alliance Railteam, established by seven most advanced HSR transport carriers in Europe, attempts were made to create such a system, having the name of Railteam Broker. Its estimated cost significantly exceeded EUR 30 million. Moreover, it turned out that its implementation would require huge changes in the process of service sales and organization of transport services in the companies, which would probably additionally increase the costs of such an undertaking. For that reason, the execution of the project was at that time abandoned [10, 11].

The strategic alliance Railteam was established in 2007. Its main objective has been to act for the increase of attractiveness and competitiveness of railways, in relation to other modes of transport, mainly by providing passengers with high standards of travelling, implementation of convenient forms of ticket purchase and seat reservation on trains, improvement of information provision, as well as various marketing activities, such as loyalty programmes. Within the framework of the concluded agreement, members of the alliance undertook the obligation to meet definite quality standards, which refer first of all to the rolling stock equipment, as well as high level of services provided to passengers. Strategic partners of Railteam were to remain independent enterprises, but through their co-operation in the alliance they were to strive for harmonizing of international high speed rail transport services in Europe [12].

At present, work has been continued in the EU to implement TAP-TSI. Those are prominent international IT projects, as yet never implemented to such a degree. Because of that it is extremely difficult to evaluate their cost, nevertheless they constitute a very important condition for the integration of national HSR systems into one coherent European system.

Benefits of High Speed Rail Integration in Europe

In the literature concerning the discussed subject, the advantages resulting from integration of transport are most often identified in three ways [13]:

- from the point of view of influence upon the environment (reduction of transport congestion, via more efficient use of available means of transport, lower levels of environmental pollution, and improved living conditions for inhabitants of highly urbanized areas, reduction of fuel consumption, enhanced efficiency of fuel use, increased safety of transport services) [14],
- from the market point of view, in the form of increased quality of services provided, reduction of costs, cutting journey/delivery time, increased availability of services, in time and space) [15],
- at the level of individual enterprises (increased demand for services, due to enhanced attractiveness of public transport, reduced own costs, through optimizing the transport offer, increased competitiveness of enterprises, access to new markets, capital, technical and organization innovations).

Integration of national high speed systems in Europe entails, first of all, increased competitiveness and attractiveness of railway in comparison with other means of transport. The cooperation between high speed service providers leads to higher quality of services provided, resulting from improved conditions of travelling, coordination of timetables from different service providers, more transparent and available information about international connections and timetables, or possibilities—for regular customers—of using special loyalty programmes. It also entails shortening the journey time and relative distance between major European urban centres.

External benefits of transport integration, that is those gained by the surroundings, including society and natural environment, are connected first of all with the changed division of transport tasks. The share of transport services provided by means of public transport increases, in the framework of which there is the use of such means of transport, which are characterized by greater transport capacity and lower external costs per passenger. The increased attractiveness and competitiveness of rail transport in Europe entails increased share of that mode of transport in the carrying of passengers and gaining the benefits related to protection of the environment, and reduction of transport congestion. Table 3 provides the average external costs of passenger transport divided into main categories (without costs of congestion) for 27 EU member states, as well as Norway and Switzerland (Cyprus and Malta have not been included in the calculations).

Rail transport is the mode of transport which generates the lowest external costs. Transferring a part of long distance passenger transport, e.g. from air transport or private cars to high speed rails leads to reduction of external costs of transport per unit of transport volume. Thus, there are measurable benefits, expressed in financial terms that can be gained by the environment and society. The change in transport execution, which would consist of replacing individual cars with rail transport, could give the benefits to society, amounting to EUR 49.3 per 1000 passenger-km

Table 3 Average external costs of passenger transport in 2008 (in EUR/1000 passenger-km) divided into main categories (without costs of congestion) for 27 EU member states, as well as Norway and Switzerland[a] [16]

Cost category	Road transport				Rail	Aviation
	Personal cars	Buses and coaches	Motorcycles and mopeds	Road transport total		
Accidents	32.3	12.3	156.6	33.6	0.6	0.5
Air pollution	5.5	6.0	11.8	5.7	2.6	0.9
Climate changes (high scenario)	17.3	9.1	11.1	16.3	1.5	46.9
Climate changes (low scenario)	3.0	1.6	1.9	2.8	0.3	8.0
Noise	1.7	1.6	14.4	2.0	1.2	1.0
Costs generated during production and distribution of energy (liquid fuels and electricity), as well as production and operation of infrastructure and rolling stock (high scenario)	5.7	2.8	3.6	5.4	8.1	7.1
Costs generated during production and distribution of energy (liquid fuels and electricity), as well as production and operation of infrastructure and rolling stock (low scenario)	3.4	1.5	2.3	3.2	3.9	3.9
Natural landscape	0.6	0.3	0.5	0.6	0.2	0.6
Biodiversity losses	0.2	0.4	0.1	0.2	0.0	0.1
Soil and water pollution	0.3	0.9	0.3	0.4	0.5	0.0
Urban effects	1.0	0.4	0.8	0.9	0.6	0.0
Total (high scenario)	64.6	33.8	199.2	65.1	15.3	57.1
Total (low scenario)	48	25	188.7	49.4	9.9	15

[a]Cyprus and Malta have not been included in the calculations

(in the scenario of high value of costs and pollution) and 38.1 EUR/1000 passenger-km in the low scenario, assuming lower emission of noxious substances. Replacement of air transport with high speed rails is a benefit for society, amounting to—on the average—5.1–41.8 EUR per 1000 passenger-km (depending on the scenario assumed).

Connecting airports that are not far from each other, by means of an HSR line (as is the case e.g. between the Charles de Gaulle (CDG) airport in Paris and the Saint-Exupéry airport in Lyon) is connected not only with closing down some of the air connections and external benefits gained from it, but also is beneficial for the airports, increasing the catchment area of those airports. By catchment areas we understand the geographic areas on which needs for transport are generated, the source of which is the requirement to get to the airport [17].

Figure 1 shows the influence of opening of an HSR connection between two airports upon their catchment areas. The red colour shows the additional area, which—due to opening the HSR connection—was added to the catchment area of both airports. As a result of providing convenient rail access to both airports, inhabitants living in the areas marked in black and red have the possibility to use transport offers of both airports [18].

Identifying the benefits resulting from integration, one should also consider the advantages resulting from a closer co-operation between high speed service providers, in the form of strategic alliances. In the literature concerning that subject, they are most often identified from the point of view of competitive forces in the M.E Porter competitive position model, that is divided into four main areas [19]:

- as concerns suppliers—greater purchasing power of alliance members enables them to conclude long term contracts on better conditions, and to negotiate lower prices,
- as concerns buyers—alliance members may offer a wider range of products and services in a bigger network, as they have a greater production potential and may offer competitive prices, thus attracting customers,
- as concerns producers of substitutive products—alliances, particularly those which refer to research and development, give the possibility of implementing modern technologies, and entering the sector of substitutive goods, at the same time,

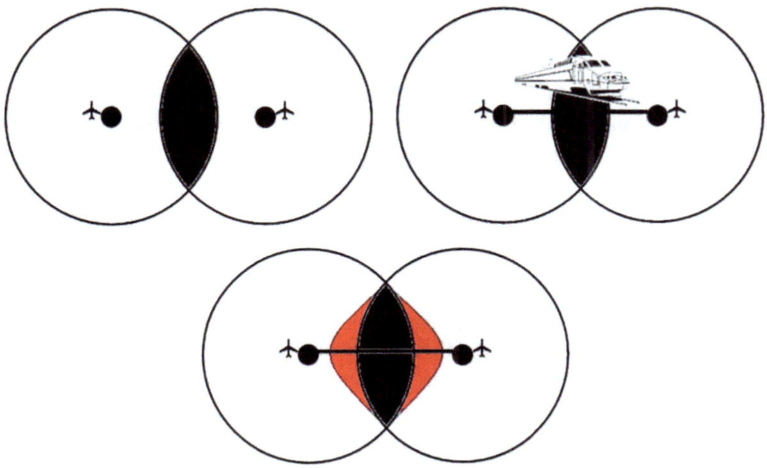

Fig. 1 The influence of opening HSR connection between airports upon the catchment areas of those airports [18]

- as regards potential new competitors—the alliance defends itself via the level of investment made, and the potential of the mother companies, experience in the sector, as well as effects of scale and field.

Durability and dynamic character of the alliance deliver fast increase of advantages for entrepreneurs which make it up, whereas the internal harmony, as well as compliance with moral principles guarantee that it does not lead to the detriment of one of the partners. A lasting improvement of the competitive position of alliance members creates—for the entire sector—a chance for improving the dynamics, extension of the life cycle, and improved profitability. Of importance is also the concern for reputation of enterprises and the entire alliance, which should allow to avoid unfair competition and reduce the costs of activities [19].

One should also take into consideration that integration may, at the same time, be a cause of such changes which may lead to deterioration of functioning efficiency of individual market entities, which is due to—among others—such phenomena as monopolization, higher entry barriers that negatively influence the strength of competitive effects and development of new bureaucratic market structures [15].

Changes that take place in transport market as a result of integration, and which are perceived as beneficial by the entities providing services on that market, are not always beneficial for the market itself. However, in case of the HSR sector, which is dominated by large state monopolies, it can be noted that integration in the form of strategic alliances is a way of obtaining access to new markets, which turned out to be a stimulator for developing competitiveness. As a result, new entities emerges, the quality of provided services increases, and transport offer becomes more comprehensive.

Conclusion

Integration of HSR, both within the same transport mode and across modes, for example with air transport, gives numerous benefits to the entities involved. The benefits are related—among others—to increased demand for services, new market entries, or reduction of risk related to investments in the development of new services. Integration in the HSR sector is also a source of social benefits, which are connected with increasing the attractiveness and competitiveness of rail transport, among other things, and the modally balanced passenger transport. As a result of integration, the temporal and spatial availability of services increases, journey time gets reduced, means of transport are used more efficiently, also the negative environmental impact and negative impact on the life quality is reduced.

Integration undertakings in the HSR sector are frequently investment projects, as they require capital- and time-intense investment in infrastructure, modern traffic management systems, IT systems assisting the seat reservation and ticket sales system, or expenditures related to the preparation, maintenance, and suitable management of contracts and agreements. The assessment of integration efficiency may be conducted with the use of standard methods used for assessing efficiency of investment projects (e.g. the net present value method, internal or economic rate of return method). This requires wide identification and assessment of costs and benefits of an integration project, in a definite time horizon, with the assessment of external costs and benefits. Unfortunately, it is an extremely labour-consuming task, mainly due to the difficulties in obtaining information and proper dimensioning of costs and benefits. What is more, integration projects are often pilot ones, in the case of high speed rail integration one can even say that in such a large international scale they are new undertakings, the costs and benefits of which are merely studied.

References

1. Directive 2008/57/EC of the European Parliament and of the Council of 17 June 2008 on the interoperability of the rail system within the Community (Recast), pp. 1-45. Official Journal of the European Union, L 191, 18.7.2008.
2. Proposal for a Regulation of the European Parliament and of the Council establishing the Connecting Europe Facility, COM (2011)665, 2011/0302 (COD). Brussel (2011).
3. Guide to COST-BENEFIT ANALYSIS of investment projects. Structural Funds, Cohesion Fund and Instrument for Pre-Accession. European Commission, Directorate General Regional Policy, pp.13-15, Brussel (2008).
4. Integration and Regulatory Structures in Public Transport. Final Report. Project Leader NEA Transport Research and Training, Study commissioned by European Commission DG TREN, pp.5-7. Rijswijk (2003).
5. Viegas, J.E.: Integrated transport systems: public – private interfaces. In: Button K.J., Henscher D.A. (eds.) Handbook of Transport Strategy, Policy, and Institutions. Handbooks in Transport. Volume 6, pp. 139 – 140. Elsevier Ltd. (2005).
6. Kos, B., Urbanek, A.: Areas of Using Telematic Tools in the High Speed Rail Integration Process in Europe. In: Mikulski, J. (ed.) TST 2014. CCIS, vol. 471, pp. 341–349. Springer, Heidelberg (2014).
7. High speed rail. Fast track to sustainable mobility, Passenger and High Speed Department, UIC, Paris 2012, p. 23, http://uic.org/IMG/pdf/uic_lgv-web.pdf [date of access: 20 March, 2015].
8. Petruccioli P.: ERTMS Benchmarking: driving costs and their actual evolution, 11th UIC ERTMS World Conference, Istanbul 1-3 April (2014), http://www.ertms-conference2014.com/assets/SESSION-PRESENTATIONS/S5/140320b-ERTMS-Benchmark -P-Petruccioli.pdf [date of access: 20 March, 2015].
9. TAP TSI Telematics Applications for Passenger Services. Technical Specifications for Interoperability. Project: TAP Phase Two Transition. 1.0-To TAP Steering Committee, 28 April 2013, Version 1.0, pp. 7. http://tap-tsi.uic.org [date of access: 20 March, 2015].
10. Barrow, K.: Harmonised fares systems are ticket to success. International Railway Journal, p. 56. December (2011).
11. Bokaie J.: Railteam aims to ground travellers. News Analysis, Marketing. p. 17, July 11 (2007).
12. Railteam Alliance, http://www.railteam.eu [date of access: 20 March, 2015].
13. Dydkowski, G.: Ocena efektywności działalności oraz przedsięwzięć rozwojowych transportu. In: Efektywność transportu w warunkach gospodarki globalnej, Michałowska M. (ed.), Prace Naukowe Uniwersytetu Ekonomicznego w Katowicach, pp. 112-117, Katowice 2012.
14. De Rus, G., Inglada, V.: Cost-benefit analysis of the high-speed train in Spain. The Annals of Regional Science 31, 175-188 (1997).

15. Michałowska, M., Tomanek, R.:, Integracja systemów transportowych jako przedmiot badań naukowych. Logistyka 2, 11-13 (2006).
16. External Costs of Transport in Europe, Updated Study for 2008, Raport, CE Delft, INFRAS, Fraunhofer ISI, Delft, pp. 70-76. Delft (2011).
17. Guidelines on State air to airports and airlines, COM 2014/C 99/03, Official Journal of the European Union C99/3, 4.4.2014, Brussel (2014).
18. Terpstra, I., Lijesen, M.G.: High-speed Train as a Feeder for Air Transport. Airlines Magazine 49, 1-5 (2011).
19. Lorange, P., Roos, J.: Strategic Alliances. Formation, Implementation and Evolution. Blackwell (1992).

Objectives and Strategies of Sustainable Urban Mobility Planning in the City of Krakow

Maciej Michnej and Tomasz Zwoliński

Abstract This chapter presents selected aspects of Sustainable Urban Mobility Planning. SUMP is a new concept of transport planning promoted by the European Commission, which in contrast to the traditional approach relies heavily on the involvement of citizens and stakeholders and institutional cooperation between various levels of government and management between neighboring municipalities. The authors provide an overview of the importance of overarching objectives and strategies of Sustainable Urban Mobility Planning for the City of Krakow.

Keywords Transport systems • Transport planning • Sustainable transport • Mobility

Introduction

It is estimated that by 2050, the urban population will increase to approx. six billion people, while nowadays about 80 % of Europeans live in urban areas. The slogan "city for people," popular among architects, begins to have a broader meaning in the context of the continuous growth of the urban population. Therefore, it is necessary to look more closely at the relationship between planning of sustainable mobility and the quality of life among residents, determined by the freedom of movement when, where, and how they want—regardless of age, income, or health status.

Krakow was the first city in Poland to adapt a sustainable transport policy in 1993 and implemented a series of programs and measures that referred to the use of space and the availability for different transport modes: pedestrian zones and zones with limited access for cars have been introduced and the infrastructure and public

M. Michnej (✉)
Faculty of Mechanical Engineering, Institute of Rail Vehicles,
Krakow University of Technology, Krakow, Poland
e-mail: maciej.michnej@mech.pk.edu.pl

T. Zwoliński
Municipality of Krakow, Krakow, Poland
e-mail: Tomasz.Zwolinski@um.krakow.pl

M. Bąk (ed.), *Transport Development Challenges in the Twenty-First Century*,
Springer Proceedings in Business and Economics,
DOI 10.1007/978-3-319-26848-4_8

transport fleets have been modernized. The city's transport policy, updated in July 2007, aims to create an efficient, safe, economical, and environmentally friendly transport system for passengers and goods.

Currently Krakow prepares to create SUMP through participating in the international project CHALLENGE—*Addressing the four Key Challenges of Sustainable Urban Mobility Planning*, the goal of which is to develop innovative, proven in practical action and possible to implement, solutions for the four key challenges in sustainable urban mobility planning [1]:

1. Stakeholder participation and citizen involvement.
2. Institutional cooperation between sectors and disciplines.
3. Identification of the most effective policy measures.
4. Monitoring and evaluation of progress in SUMP development.

For each abovementioned challenge, the cities analyze their local mobility situation, develop new strategies on how to tackle their urban mobility problems, and test solutions to overcome barriers in participation, cooperation, measure selection as well as monitoring and evaluation in more than forty pilot schemes [1].

Current Transport Situation in Krakow

Krakow is the second largest city in the country, located in southern Poland with ca. 760,000 inhabitants. Its unique historical, cultural, and scientific character still dominates the city's economic and tourism development. Krakow hosts around eight million tourists each year, and the Jagiellonian University, with some 140,000 students, is one of the oldest universities in Central and Eastern Europe (Fig. 1).

Krakow is known for its well-developed public transport system. Nowadays, there are 132 bus lines and 24 tram lines serving ca. one million passengers daily. The quality of rolling stock and infrastructure is being improved constantly, including offering, for example, modern ticketing solutions and demand responsive bus transport.

Congestion in Krakow has significantly risen in the past decades and reached an unacceptable level. High traffic volumes are observed in the city center as well as on the two ring roads. The entire region shows high congestion levels as well. It has been observed that rush hours have prolonged up to 5 h in mornings and afternoons which is a considerable rise in peak hours when traffic is at its highest.

The increased motorization rate is considered to be one of the most important causes of congestion problems. The motorization rate in Krakow (the total number of vehicles per 1000 residents) is comparably high in the European context (644 vehicles per 1000 residents in 2012) (Table 1).

Parking standards (e.g., parking-space-to-unit or -inhabitant ratios) are set in local development plans but only about a third of the city's area is covered by such local plans; implementing parking requirements in the remaining areas is not regulated in a uniform way.

Krakow is one of the most polluted cities in Europe and meets neither European nor national air quality standards. It was reported that air quality standards are

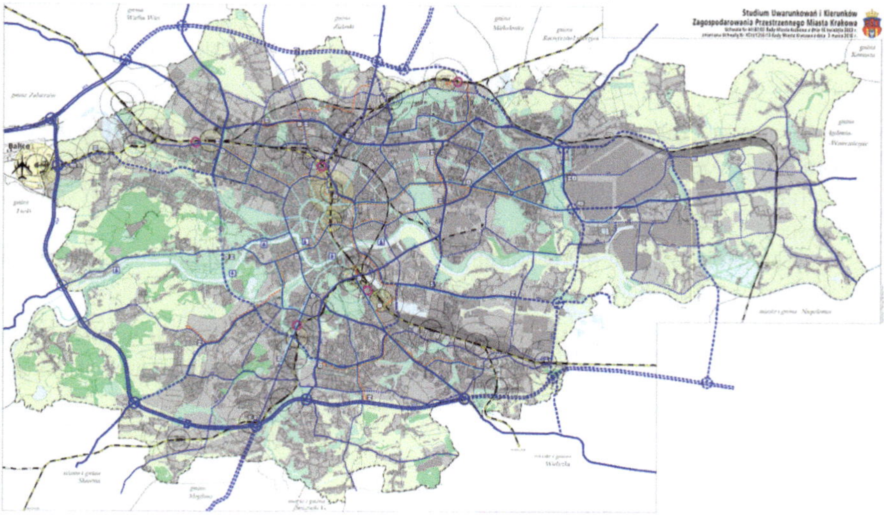

Fig. 1 Map of Krakow and its transport system [4]

Table 1 Transport data/modal shares in Krakow

Mode	Share in trips (%)
Walking	28.5
Private car	30.3
Public transport	39.1
Cycling	2.1

Survey 2013/14 [4]

exceeded approx. 200 days a year. For example, a study conducted in 2012 [2] suggested that the PM2.5 (PM—particulate matter; fine particles that have diameter of 2.5 µm or less) and PM10 (coarse particles fraction in addition to the PM2.5 fraction) levels were exceeded in all measure points in Krakow. Also the average NO_2 (nitrogen dioxide) emissions measured at a station on the second ring road were 80 % over the allowed standard.

SUMP Objectives and Strategies

Sustainable Mobility Planning Background in Krakow

As yet, Krakow has not published the full SUMP document. There are many sectorial documents, and the ones which are most significant for the local transport are: the transport policy (2007), the parking program (2012), and the so-called Transport Plan (2013). The latter deals primarily with the development of public transport in the city and region. All these documents will be the basis for the forthcoming SUMP development. Currently there are three main objectives for the city in terms of development of the transport system:

- To ensure the efficiency of the transport system of the city as sustainable in economic, environmental, and social terms;
- To further strengthen the role and improvements in the quality of public transport;
- To create an integrated metropolitan transport system, providing accessibility in the framework of the region, the country, and Europe.

The time horizon of the SUMP will be the year 2025 or 2030. It is planned to update the transport policy within 2015, and use it as a basic vision-establishing document for the SUMP purposes.

SUMP Objectives for Krakow

The following table provides an overview of the importance of overarching transport planning objectives for Krakow (Table 2).

The application area of objectives presented in table 2 covers the whole City of Krakow, and includes surrounding municipalities—especially in terms of public transport (buses) and rail connections with Krakow. As described above, Krakow aims to create the integrated transport system also in terms of the national and European needs (i.e., through the existing and modernized/extended airport), and therefore, already existing planning documents take this important objective into account.

SUMP Strategies for Krakow

The strategy is a response to the fundamental question: "how to fulfill the objectives." Strategies for SUMP in Krakow relate to all forms of mobility—both collective transport, individual, freight, bicycle, pedestrian, parking policy, and road safety. According to current challenges in urban mobility in Krakow, the main strategies include:

- Reducing car use—by constantly implementing access restrictions measures—The City of Krakow has already taken several actions to tackle congestion caused by cars, public transport, and freight. A number of demand management measures have been introduced such as the introduction of parking restrictions (fee-based) and access restrictions separating the city in zones A, B, C; entering zones A and B is only allowed for inhabitants and for delivery vehicles during certain spans of time. The creation of new links between districts by building new ring roads

Table 2 Description of sustainable mobility planning objectives for Krakow

Objective	Description and potential targets
Efficiency	Maximizing the benefits which users can gain from the transport system, after taking into account the costs of provision and operation of the system Potential targets: To improve the energy efficiency of vehicles in all modes of transport To introduce and develop fuels and propulsion systems compatible with the principle of sustainable development To promote efficient, innovative, and multimodal transport services To maximize the efficiency of infrastructure use (toll systems, ITS, capacity improvement programs) To establish ecological, high mobility corridors To provide a framework for specific revenue generated by transportation
Livable streets	Pleasant street and outdoor conditions in residential areas. It includes the positive external effects on social, cultural, and recreational activity in neighborhoods, freedom of movement on foot and bicycle, and reduced sense of danger for these modes. It is linked to, but separate from, the environmental and safety objectives Potential targets: To develop the transport system without barriers for pedestrians and cycling To maintain the comfortable density of pedestrian crossings, ensuring adequate width of sidewalks and crosswalks To extend the range of zones with traffic calming To introduce restrictions for heavy goods vehicles in the city center and selected residential areas To coordinate spatial development policy in Krakow and neighboring municipalities in order to properly direct the process of devolution settlement, having the effect of reducing traffic and transport, especially the car and realized counters the undesirable dependency on this means of transport To provide preference for settlement development in the peripheral zones and suburban areas, which are conveniently linked by rail or tram to areas of concentration of jobs and services in Krakow To create or transform the structure of housing units and service in a friendly movement pedestrian, bicycle and people with disabilities and the focus on the development of public transport services, including the provision of convenient and safe walk to schools and bus stops

(continued)

Table 2 (continued)

Objective	Description and potential targets
Environment	The environmental impacts concerning transport include noise, air pollution, vibration, visual intrusion, severance, fear and intimidation, and the loss of flora and fauna, ancient monuments and historic buildings through the consumption of land. The environmental protection objective involves reducing the impact of transport facilities, and their use, on the environment of both users and nonusers Potential targets: To replace the bus fleet with the one that meets European emissions standards To implement the "communication zones with reduced emissions" in order to eliminate the vehicles that do not comply with the emission standards To introduce weighing stations for moving vehicles—elimination of overloaded vehicles that damage road surfaces To develop "user-friendly" public transport To construct bicycle paths To ensure public transport priority To improve the parking policy (e.g., "Park and ride") To make the land use planning take into account noise pollution—zoning of building functions (service and commercial buildings closer to the traffic, housing-on) To promote effective and consistent enforcement of the access restrictions, speed limits, tonnage limits To implement monitoring through a system of noise management
Equity and social inclusion	Equality, especially between different groups in society, in opportunities to travel, costs of travel, and environmental and safety impacts of travel Potential targets: To increase the participation and involvement of citizens and NGOs in the development of transport in Krakow and setting the quality mobility To develop social dialogue between the Municipality and the citizens and nongovernmental organizations To develop the participation of NGOs in the joint creation of diagnoses, the exchange of information between the NGO sector and the Municipality To activate participation in the public consultation and improve the quality of consultation To support the creation, operation, and recognition of institutions of civil dialogue and cooperation To increase the participation of NGOs in the evaluation of public tasks To develop a support system for initiatives/projects implemented by NGOs

(continued)

Table 2 (continued)

Objective	Description and potential targets
Safety	Minimizing the number of all types of road traffic accidents. Usually expressed through total traffic accident costs or by accident risk per vehicle kilometer Potential targets: To improve the condition and security of the residents To update the rights of passengers To generate universal application and enforcement of standards; safety, security, environmental protection, and working conditions in transport To improve the security control methods in order to ensure a high level of protection with the least inconvenience to passengers
Economic growth	The increase in the market value of the goods and services produced by economy over time Potential targets: To increase the economic efficiency and operational efficiency of public transport using the principles of market economy and competition To effectively implement economic and financial policy To develop methods for the analysis of economic and financial performance and functional development of possible scenarios for the transport system To create friendly environment for investors To encourage effective cooperation of the city government, the business community and R&D institutions
Finance	The procurement (to get, obtain) of funds and effective (properly planned) utilization of funds. It also deals with profits that adequately compensate for the cost and risks borne by the transport Potential targets: To intensify acquisition of external funds for the development of the transport system, actively raising funds from the European Union and the creation of conditions for using those funds To attract private capital to public investments (public–private partnership)

also aims to ease the congestion. However, road infrastructure development might be seen contradictory to sustainable mobility and might stimulate rather than reduce the car-use.

- Improving the use of road space—by changing the use of existing road space, in terms of freeing the space to pedestrians and cycling, closing the roads, including them in the enlarged pedestrian zones.
- Improving the use of public transport—Significant public transport improvements were undertaken in the past ten years in Krakow, which not only led to the effective public transport system but also created its positive image. According to independent studies, Krakow's public transport is the most effective one in Poland (based on various data and indicators) [6]. Reliability and punctuality of public transport is high; in addition, regular surveys of customers conducted by the local public trans-

port operator indicate high satisfaction rates that steadily increase. The number of public transport passengers is relatively stable; no major shifts to other transport modes, cars in particular, have been observed. For the most part, public transport is run by the public transport operator MPK which has the share of about 90 % of the lines. Moreover, Krakow is served by one private transport operator. Since January 2013 the tram network has been put in focus as the bus system is now seen rather as a complementary service. This controversial decision was not in favor of all local agents. Integrated ticketing is available for all services including tram, bus, and national rail in the region. Unchanged for many years, the public transport fares have been recently raised. The fares are now higher than in other Polish cities, which is widely accepted due to the good fare-to-quality ratio.
- Improving walking and cycling—there is a program of extension of bike paths network and other measures (like contra-flow lanes, tempo 30 zones, etc.) are in place, which is particularly prominent for the development of cycling.

Conclusions

Urban transport and mobility should not be treated as a goal in itself. It should have a positive impact on higher objectives, such as quality of life and well-being of citizens [3, 5]. This is the starting point for the concept of sustainable urban mobility planning (SUMP). Since late 1980s, the city of Krakow has developed and introduced a number of sustainable urban transport measures, especially those concerning access and parking restrictions and the network of high quality public transport. Recently, there is a huge and promising interest concerning the potential for cycling in Krakow, including further development of bike rental scheme. All these actions confirm the willingness to at least keep the balance between motorized and non-motorized modes of transport in Krakow. However, the full SUMP planning concept needs to be introduced in order to improve aspects of citizens involvement and institutional cooperation. SUMP shall also help to improve active involvement of neighboring municipalities, as it is more and more important to consider transport planning for Krakow as a metropolitan challenge.

References

1. Annex 1 – Description of Work, Addressing Key Challenges of Sustainable Urban Mobility Planning, Acronym of the project: CH4LLENGE, kontrakt nr IEE/12/696/SI2.644740.
2. Assessment of air quality in Lesser Poland district. Regional Inspectorate of Environment Protection. Krakow, 2013.
3. "GUIDELINES - Developing and Implementing a Sustainable Urban Mobility Plan"- Rupprecht Consult – Forschung und Beratung GmbH Contract ELTISPLUS, nr EACI/IEE/2009/05/S12.558822, www.mobilityplans.eu.
4. Report Deloitte/Millward Brown SMG/KRC, zikit.krakow.pl.
5. "The state-of-the-art of sustainable urban mobility plans in Europe" September 2012, Rupprecht Consult and Edinburgh Napier University, www.mobilityplans.eu.

The Significance of Pedestrian Mobility as Shown by the Example of the City of Gdynia

Marcin Wołek

Abstract The specificity of the cities and the dynamics of the various processes which occur within them result in conflicts within the urban space, the effect of which is, among others, suburbanisation bearing negative consequences on sustainable transport and urban mobility. The analysis of EU policies has served as a basis for determining the increasing role of pedestrians in shaping of sustainable mobility policies.

The significance of pedestrian mobility in city centres is discussed using the example of the city of Gdynia. The articles present findings of research into the volume and the structure of traffic on one of the main roads in the centre of Gdynia. The challenges which need to be addressed when carrying out of complex research into pedestrian mobility have also been determined.

Keywords Pedestrian • Mobility • Modal split

Introduction

Every city constitutes a complex organism of a compound and dynamic structure. The relationship between every city and its environment can be characterised by variability. Particular functional subsystems of a city strive to achieve balance; however, they never succeed. A city, as a whole, remains in a state of a permanent imbalance, the source of which should be seen in both external and internal factors. The well-known saying about climate, which states that its only stable and foreseeable feature is its variability, can also be easily used to describe cities. Their variability was discussed by J. Regulski, who claimed that the fact that their development is a constant process results from its particular participants (people, households, companies and other organisations) striving to maximise either economic benefits or satisfaction. He also stated that such complex systems as cities cannot achieve a state of

M. Wołek (✉)
Chair of Transportation Market, University of Gdańsk, Sopot, Poland
e-mail: mwol@wp.pl

M. Bąk (ed.), *Transport Development Challenges in the Twenty-First Century*, Springer Proceedings in Business and Economics,
DOI 10.1007/978-3-319-26848-4_9

balance due to mutual relationships between the activities of subjects that function within their space and that they shall remain in a perpetual state of evolution [1, p. 93]. A similar thesis was put forward by K. Button who wrote that "cities never were and never shall be <<static units>>" [2, p. 217]. The definition of a city's mega-product proposed by T. Markowski describes it as a "balance of external effects" [3, p. 224] and also refers to the conflicting interests of different clients—its users.

The role of cities in the civilisational development of humanity requires no further discussion. All centres of this development that existed up until present times in human history were concentrated around cities, although the main driving forces behind their development differed (religion, defences, trade, etc.). "Before the Industrial Revolution the spatial structure of cities was determined by the needs of pedestrians" [4, p. 9]. Currently the role of cities is strengthened by globalisation trends as well as by technical development together with the continual growth of urbanisation in the world. The prominent role of cities is also recognised by global market business entities, i.e. one of the development scenarios of Shell company (plan known as "Mountains") emphasises the role of authorities in the development of "compact cities" and the transformation of global transport network [5, p. 125]. In turn, among the six megatrends ("megatrends are large, transformative global forces that impact everyone on the planet. EY has identified six megatrends that define our future by having a far-reaching impact on business, society, culture, economies and individuals") identified by the most recent report of Ernst & Young [6, p. 2–3] were:

- Digital future
- Entrepreneurship rising
- Global marketplace
- Urban world
- Resourceful planet
- Health reimagined

The urban world megatrend is closely related to such megatrends as digital future, entrepreneurship rising and resourceful planet. The report also defines three features of the city of the future, namely, competitiveness, sustainability and vitality [6, p. 31]. The latter in a more narrow context stands for the possibility to decrease the dependency of the cities on fossil fuels in an acceptable and feasible way in a social and economic respect [7, p. 6].

Specificity of Transport Services in City Centres

An important practical and theoretical issue of the economics of a city is the connections of its centre (central area) with other areas. City centres constitute the oldest, historically shaped tissue, which bears a significant meaning for the image of cities and the identity of their inhabitants. They offer, however, limited possibilities for intervention as the residential, office, service, tourist and recreational functions are interwoven. These are the generators of mobility on a local, regional and even global scale. The accumulation of functions, coupled with the willingness to provide an appropriate accessibility to all "users" of the city, constitutes a source of unceasing conflict of economic, spatial and environmental, as well as social, dimensions.

The development of public transport followed by the access of individuals to motorised transport, together with the road infrastructure, enabled the process of spatial expansion of cities. A few phases of this process can be distinguished among which that of urban sprawl is identified as suburbanisation. The main factors which influence the dynamics of this process are the differences in the price of properties located within the cities and those located in their vicinity, where the commute cost from the suburbs to the centre is relatively small. These are in fact external factors in relation to the city affected by the process. The poor quality of urban space (including buildings), low attractiveness of the local job market or the level of public service may all constitute external factors. The literature on the economics of American cities names factors influencing the suburbanisation process as poor quality of a city's accommodation, avoidance of racial conflicts and the lack of willingness to live close to poorer districts, tax policy of a city, crime as well as education [8, p. 207–208]. 1/3 of the total number of trips connected with commuting to work is fulfilled within the city centre whilst 20 % on the route between the suburbs and the centre [8, p. 287].

The answer to the crisis of the central area of a city is given in the form of a complex processes of revitalisation, which should be carried out in technical (modernisation and construction of technical and social infrastructure) and social (involvement of local communities, activities aimed at the improvement of the quality of local services, introduction of new functions, i.e. business and social) dimensions. In respect of the transport dimension, the revitalisation of city centres is tantamount to the increased role of pedestrians in the core centre of a city, a convenient commute via public transportation and cycling, as well as a reduction in accessibility for cars. To sum up, the processes of revitalisation pose a chance to completely restructure the transport system of a city, what in turn leads to a better balance in the modal split and a decrease in the arduousness of transportation for inhabitants. These processes fall in line with the suggestions of the European Commission with regards to mobility in urban areas.

Sustainable Urban Mobility as an Element of EU Transport Policy

The Urban Mobility Package from the end of 2013 expanded on the problems presented in the White Book 2011, whilst at the same time attempting to horizontally integrate and synthesise some of the issues. Mobility within urban areas of the EU is responsible for 23 % of the total CO_2 emissions.

From the point of view of the functioning of a city, the most important challenge is posed by transport congestion and related costs, the burden of which is inflicted upon all customers of a given urban system. There is limited potential for the reduction of the negative effects resulting from transport congestion and the level of influence of the European Commission. It largely falls to local authorities to reduce and counteract these effects. Therefore, the Commission's aforementioned efforts create an additional mechanism encouraging local authorities to take on more complex initiatives connected with the shaping of urban mobility. "Local authorities aim to discard the past, isolated approach and to create strategies which could serve as an incentive to change to more eco-friendly and sustainable modes of transportation such as walking, cycling, or using public transportation, as well as to take on new models of utilization and ownership of vehicles" [9, p. 3–4].

One of the leading trends within sustainable urban mobility is the increasing role of pedestrian mobility, especially within the central areas of cities. The increase of the share of pedestrians in a city's modal split is deemed to be one of the solutions which is most effective economically, socially and environmentally. It has to be said, however, that a precise evaluation of the size of the share pedestrian mobility has in the modal split presents serious methodological and organisational difficulties. The sole definition of a pedestrian trip is not unambiguous and can be understood differently depending on the particular country and established examination procedure. To give an example, in the most recent research into transport preferences and behaviour carried out in 2015 among the inhabitants of Gdynia, a "pedestrian trip" is defined as a relocation of at least 500 m inclusive of reaching any mechanical modes of transportation.

Moreover, pedestrian relocations accompany practically every single way of travel fulfilled by mechanical modes of transportation (by car, bicycle, public transport). They are usually characterised by short distances. Despite the fact that the share in the number of trips may be generally high, in the dimension of passenger kilometres, it is significantly lower than in the case of trips fulfilled by mechanical modes of transportation.

The above issues should not, however, obscure the benefits that result from the increase of pedestrian traffic in urbanised areas. The development of pedestrianisation processes (the increase in the attractiveness of spaces for pedestrians and promotion of pedestrian travel) is accompanied by the popularisation of cycling. It is necessary to make consequent decisions in the area of spatial planning, investments in infrastructure as well as the organisation of individual and public transport for this increase to be long term and permanent [10, p. 5]. Such decisions, due to their strategic character, investment outlays and changes that they bring about for the inhabitants, should be based on wide-scale primary research taking into account the complex chain of mobility of all "users" of a city's space.

Pedestrian Relocations in Gdynia City Centre in 2014 in the Light of the CIVITAS DYN@MO Project Research

As part of the CIVITAS DYN@MO project, activities are undertaken in Gdynia in order to increase the share of pedestrians in the modal split. They are accompanied by large-scale research into pedestrian activity that until recently was only fragmentary. Research conducted as a part of the CIVITAS DYN@MO project in May 2014 included three streets of Gdynia, namely, Świętojańska, Starowiejska and Skwer Kościuszki. Measurements were inclusive of all road vehicles, including bicycles. They also took into account the volumes of passengers on board public transport vehicles and the number of pedestrians entering the section. The methodology of research involved observation by students and employees of the Faculty of Economics. This article presents the results for Świętojańska Street (section marked by the bus stop opposite Gdynia Infobox) on a typical working day (between 5 am and 10 pm). Figure 1 presents the localisation of the section.

Świętojańska Street is one of the most important streets in Gdynia. Leaving aside its historical value, symbolism, localisation and multifunctionalism, it constitutes the basic element of the transportation layout of the centre of Gdynia. It was revitalised as a part of the CIVITAS TELLUS project (2001–2006). The aim of the revitalisation was to increase the space designated for pedestrians, to organise parking and to carry out a major aesthetic makeover of the space (pavement surface, small architecture, lighting, urban greenery, modernisation of the traction network for trolleybuses). The introduction of a pay and display parking zone in the city centre in 2008 constituted the next step in ordering the transport network on Świętojańska Street. It organised parking and allowed for the rationalisation of the utilisation of space in Gdynia city centre, given the dynamic increase in the access of individuals to motorised transportation.

The average volume was taken into account when determining the number of car users. According to the recommendations for car transportation put forward by the JASPERS initiative, it varies between 1.2 and 1.6—depending on the type of travel. The lowest average indicators of volume are attributed to the work commute (motivation home—work—home) and business travel. The highest is ascribed to other

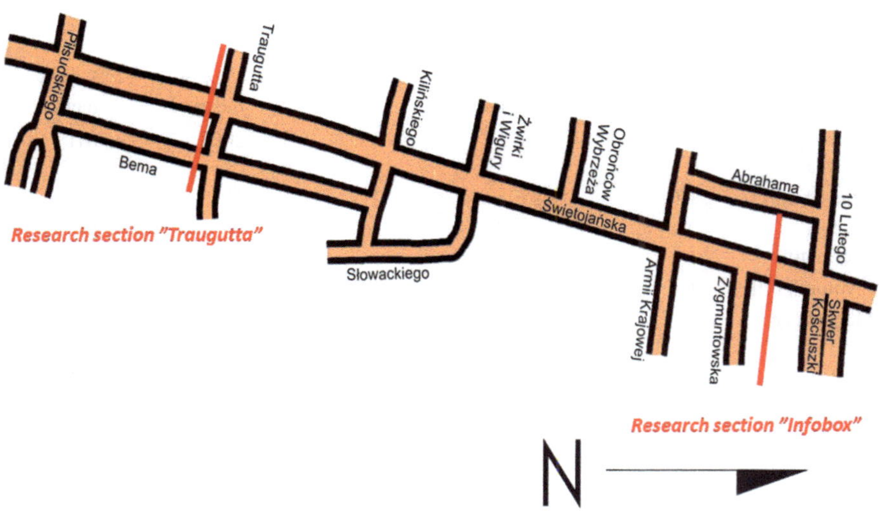

Fig. 1 Localisation of the measurement sections on Świętojańska street in Gdynia (29.05.2014) (source: own study)

types of travel [11, p. 34]. The following proportions were assumed for trips fulfilled by car: work commute 15 %, business 10 % and others 75 %. This means that the average general volume of passengers in a car amounted to 1.5 persons per vehicle.

For Świętojańska Street in the examined section, cars were the main mode used for relocations with nearly a half of noted trips using this mode of transportation (with the assumption that each vehicle was carrying 1.5 persons on average). Trolleybuses are also an important mode of transportation. They made up at least ¼ of the share of transportation in the modal split in general trips in all sections included in the research. Pedestrian traffic also proved to be an important component of relocations in the aforementioned section. Its share was nearly the same in both directions and amounted to between 23 and 24 % (Fig. 2). This result affirms the significance of pedestrian relocation in central areas of cities and is approximate to the results of research conducted in other European cities of a similar number of inhabitants in which the share of pedestrian travel amounts to between 18 and 28 % (Table 1).

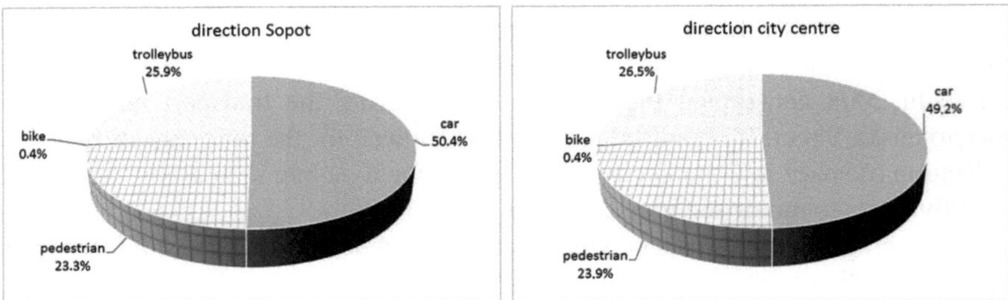

Fig. 2 Weighted traffic structure on Świętojańska Street section (Infobox) in both directions on 29.05.2014 [10]

Table 1 Pedestrian share in the modal split of selected European cities [12]

	Pedestrian share in the modal split (%)	Population [thous.]	Density [inhab./sq. km]	Year of measure	Survey method
Bonn (Germany)	28	327.9	2322	2008	RHS
Graz (Austria)	19	270	2117	2013	RHS
Aachen (Germany)	23	258.7	1608	2011	HC
Plymouth (UK)	18	256	3207	2010	PSPS
Magdeburg (Germ.)	21	234.8	1164	2008	RHS
Saarbrucken (Germ.)	23	177.2	1061	2010	RHS
Szeged (Hungary)	22	161.8	576	2009	HS

RHS representative household survey, *HC* household census, *PSPS* public satisfaction postal survey, *HS* household survey

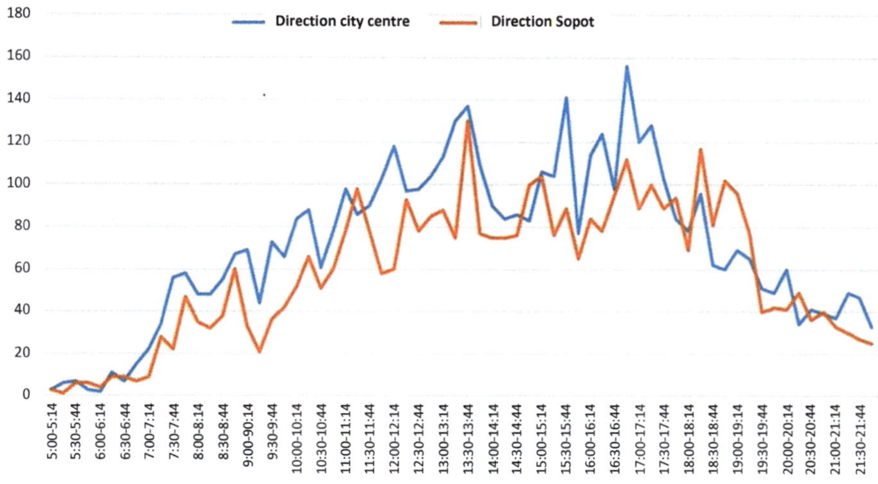

Fig. 3 Volume of pedestrian traffic within a section of Świętojańska Street in the vicinity of Infobox on 29.05.2014 [15 min. time slots] [12]

The analysis of the volume of pedestrian traffic in a period of 24 h points to a relative balance between the directions of pedestrian travel in the analysed section (Fig. 3) in the first afternoon rush hour that falls between 1 and 2 pm. The second rush hour, which falls between 4.30 and 5.30 pm, is characterised by a distinct advantage of pedestrians travelling into the city centre. The decrease in pedestrian traffic volumes in the same direction after 6 pm has to be emphasised as it highlights the strong, prevailing service and commercial functions of this street.

Volume of pedestrian traffic presented below might be treated as a demand for space in given urban area and depends on various internal and external factors.

Summary

The increase in significance of pedestrians in spatial planning and their influence on sustainable urban mobility suggest that broader research into pedestrian traffic should be carried out, especially in connection with central areas within cities. The knowledge of pedestrian relocation proves to be useful and it goes beyond the sphere of spatial planning and mobility. Institutions and businesses (mainly commercial) constitute potential beneficiaries.

Taking into account pedestrians in the modal split requires reliable research findings that should be characterised by regularity, complexity and their ability to be compared in regards to time and space. The increase in the significance of pedestrian relocation should constitute a vital component of revitalisation projects carried out in city centres. It is required that research into pedestrian traffic be continued and expanded by including such quality indicators as the trip's truss, motivation for undertaking it and its duration. The argument is strengthened by the fact that first attempts at the implementation of pedestrian traffic into urban traffic models are being undertaken.

The presented findings of the 2014 research conducted on Świętojańska Street in Gdynia constitute one of the elements in the creation of a complex information system on sustainable mobility. They highlight the high share of pedestrian relocation in the modal split in the city centre. This share is comparable to other European cities and amounts to between 23 and 24 % on a typical working day. The observation method adopted in the research allows for an examination of traffic volumes in both directions as well as of the modal split between the basic means of transportation within the city. In order to obtain qualitative data, it is required that other ways of communication with the respondents be employed which results in the research being a more costly and lengthy process.

References

1. Regulski J.: Elementy teorii polityki urbanistycznej. [W:] Miasto i jego władze. Studia nad ekonomiczną teorią rozwoju miast. Red. J. Regulski. Zakład Narodowy im. Ossolińskich, Wrocław 1984
2. K. Button: City management and urban environment indicators. "Ecological Economist" 2002 nr 40
3. T. Markowski: Zarządzanie rozwojem miast. Wyd. PWN, Warszawa 1999
4. The Trolleybus as an Urban Means of Transport in the Light of the Trolley Project. Edited by M. Wołek and O. Wyszomirski. Uniwersytet Gdanski, Gdansk 2013
5. A. Wilkinson, R. Kupers: Living in the Futures: How scenario planning changed corporate strategy. "Harvard Business Review" 2013 May 2013
6. Megatrends 2015. Making sense in a world in motion. Ernst & Young 2015
7. P. Newman T. Beatley, H. Boyer: Resilient Cities. Responding to Peak Oil and Climate Change. Island Press, Washington-Covelo-London 2009
8. A. O'Sullivan: Urban Economics. 5th edition. McGraw-Hill, New York 2003
9. Together towards competitive and resource-efficient urban mobility". Communication From The Commission To The European Parliament, The Council, The European Economic And Social Committee And The Committee Of The Regions. Brussels, dnia 17.12.2013 r. COM(2013) 913 final
10. M. Wołek: Badania natężenia ruchu w centrum Gdyni. raport opracowany w ramach projektu CIVITAS DYN@MO, Gdynia, czerwiec-wrzesień 2014
11. Infrastruktura drogowa. Niebieska Księga. JASPERS (Joint Assistance to Support Projects in European Regions), Warszawa 2008
12. EPOMM data base: http://www.epomm.eu

Cars and Urban Travel

Katarzyna Hebel

Abstract The present article is an attempt to describe the role of cars in the urban transport of inhabitants living in select European cities. It uses information obtained from research into modal split that has been conducted throughout the last 10 years. This allowed for trends of change to be revealed with reference to the cities in which research was conducted on a number of occasions throughout the aforementioned period.

The role of cars in the undertaking of urban travel with regard to the Polish cities was determined on the basis of findings of complex traffic research as well as preferences and travel behaviour research. The research was conducted in select Polish cities. The factors determining the decision to undertake urban travel by car were also established.

Keywords Car • Urban transport • Travel behaviour

Introduction

The role of cars in urban travel is systematically on the increase. At the same time, the attitude of society towards cars is changing, as it ceases to be a symbol of affluence and elevated social status and instead becomes a common consumer commodity, virtually indispensable in everyday life. It seems impossible to reverse the trend of increasing motorisation and wide use of private cars in urban travel which has lead to congestions (traffic) in a growing number of cities. The fact that a car permits its passengers to travel door-to-door means that it provides freedom of travel within cities only until infrastructural factors, such as the quality and parameters of the roads and parking spaces, cause congestion and result in a decrease in the attractiveness of cars as a mode of transportation.

Research into the ways in which trips are fulfilled serves as a means for establishing the actual significance of cars. Such research is conducted by employing

K. Hebel (✉)
Chair of Transportation Market, University of Gdańsk, Sopot, Poland
e-mail: khzh@wp.pl

M. Bąk (ed.), *Transport Development Challenges in the Twenty-First Century*,
Springer Proceedings in Business and Economics,
DOI 10.1007/978-3-319-26848-4_10

various methods and techniques, thus findings have to be compared carefully. They allow, nevertheless, the trend of change to be verified and serve as a basis to determine the key moment at which to undertake actions aimed at decreasing the negative consequences arising from the overwhelming share of cars in urban travel.

Cars and Urban Travel in European Cities

The findings regarding the ways in which urban trips are undertaken are stored on the European Platform on Mobility Management. The platform offers access to research findings from the past 10 years (2004–2014) which were gathered in 430 cities (in some of them, research was conducted a number of times). In 60 % of the cities (260 cities out of 430) for which data was gathered, over 50 % of urban trips were undertaken by car. Lone drivers made up nearly 100 % of those travelling during such trips.

Research findings in large cities (above 500 k inhabitants) of the most current period, namely, the years 2010–2014, served as a basis for the illustration of the role cars play within cities (Fig. 1). This allowed the relevance and clarity of the data presented to be preserved.

The graph presents the cities in decreasing order according to the share of cars in urban trips in the modal split structure. The largest share belongs to car trips in the cities of the United Kingdom and Ireland, such as Manchester, Leeds, Sheffield and Dublin. Wrocław, Poznań and Cracow were among the European cities in which cars are used extensively in urban travel.

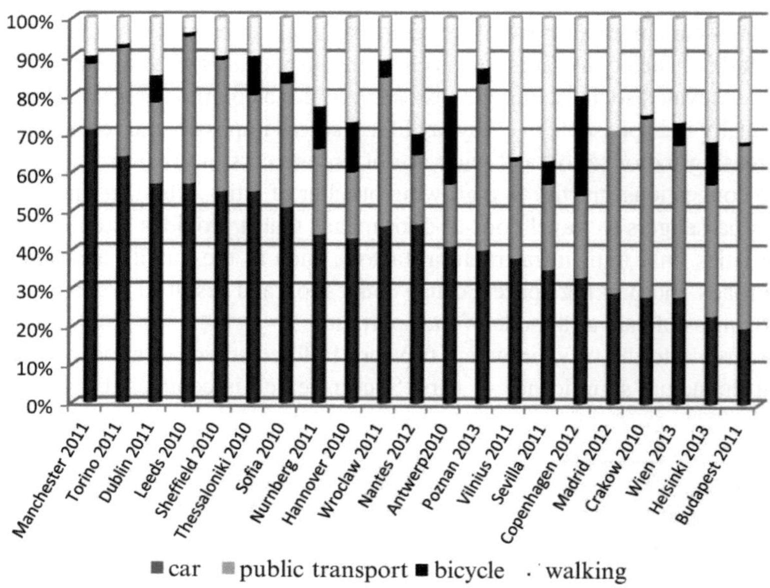

Fig. 1 Modal split of urban trips in select European cities (number of inhabitants above 500 k) in the years 2010–2014 in the light of the research [3]

Cars and Urban Travel in Chosen Polish Cities in the Light of Complex Traffic Research

The way in which inhabitants undertake urban travel in different cities suggests strong dependence on local conditions. It was established on the basis of complex traffic research (CTR), which was fulfilled between the years 2004 and 2014 in select Polish cities (Warsaw 2005, Tychy 2008, Wrocław, Rzeszów and Gdańsk 2009, Szczecin 2010, Wrocław 2010/2011, Cracow 2013), that the inhabitants of Warsaw and its vicinity fulfilled 30 % of their trips undertaken in the year 2005 by car (Fig. 2) and 35 % by inhabitants of Gdansk in 2009, and the inhabitants of Wrocław in the period 2010/2011 chose this mode of transportation for 43 % of their trips (Fig. 3).

It is interesting to note that in Cracow in 2013, it was established that inhabitants chose to travel by car during only 34 % of their trips [1, pp. 32–33]. Among the drivers who took part in research, as much as 61 % of them were travelling alone with no passengers in the car. In 28 % of cases, there was a single passenger

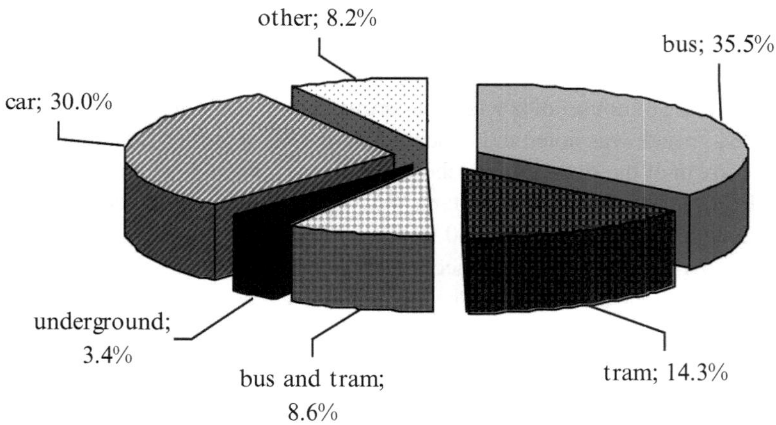

Fig. 2 The mode of undertaking urban travel by Warsaw inhabitants in 2005 [4]

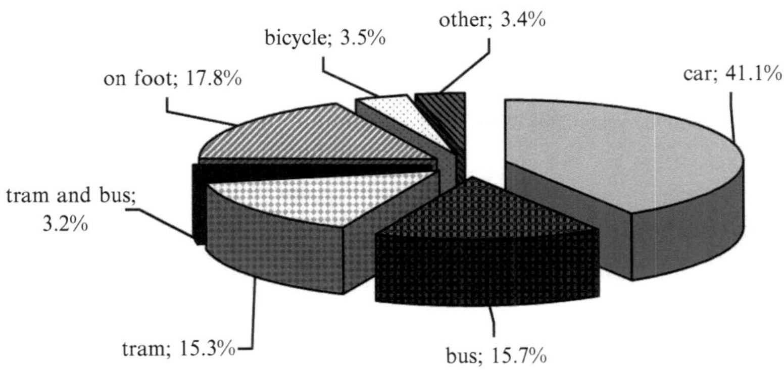

Fig. 3 The mode of undertaking urban travel by Wrocław inhabitants in the period 2010/2011 [5, 6]

travelling and in 7.5 % two passengers. In the very centre of Cracow, where traffic is restricted and parking charges are introduced, cars are chosen by even fewer inhabitants and their share in the modal split amounts to as little as 27 % [2, pp. 30–34]. This serves to confirm the significance of local conditions in the decision to undertake urban travel by car.

Cars and Urban Travel in Gdynia in the Light of Research into Transport Preferences and Behaviour

In a few Polish cities, research into the transport behaviour of their inhabitants is conducted systematically with the use of the same methodology. In Gdynia such research was conducted as part of research into transport preferences and behaviour of the inhabitants in the years 2004, 2006, 2008, 2010 and 2013. On each occasion a sample of 1 % of inhabitants was surveyed. Random sampling was carried out maintaining its correct structure with regard to age, gender and district. Household research was conducted through use of a questionnaire.

Households in Gdynia with a car constitute a growing group. Throughout the research period, the share of households without a car dropped from 47 to 26 %, whilst the share of households with a car grew from 53 to 74 %. It is noteworthy that the largest growth was noted in households with only one car (Table 1).

The growth of the share of households with cars reflected the growth of main car users (Fig. 4). The share of inhabitants who are not main car users in their households remained at approximately 30 %.

In keeping with the increased access of the inhabitants of Gdynia to cars, their transport behaviour is changing. The car is used increasingly often in urban travel. The research conducted in Gdynia determined this fact by firstly asking respondents to declare their habitual mode of travel to later verify the answer by enquiring about the actual way they travelled on the day prior to the survey.

Among all surveyed inhabitants of Gdynia, the group that declares that it travels around the city by car is increasing (always, nearly always and nearly as much as by public transportation). The share of those who always travel by car has seen the most growth. Throughout the entire research period, their share increased by 2.5 times (Table 2).

Table 1 The structure of Gdynia inhabitants according to their motorisation status in the light of market research in the years 2004–2014 [%] [7–11]

	Year				
Motorisation status	2004	2006	2008	2010	2013
Inhabitants from households with no cars	47.37	36.71	35.91	31.75	26.35
Inhabitants from households with one car	45.44	51.19	52.26	54.38	57.05
Inhabitants from households with two cars	6.43	10.35	10.49	12.10	15.00
Inhabitants from households with three or more cars	0.76	1.75	1.34	1.77	1.60

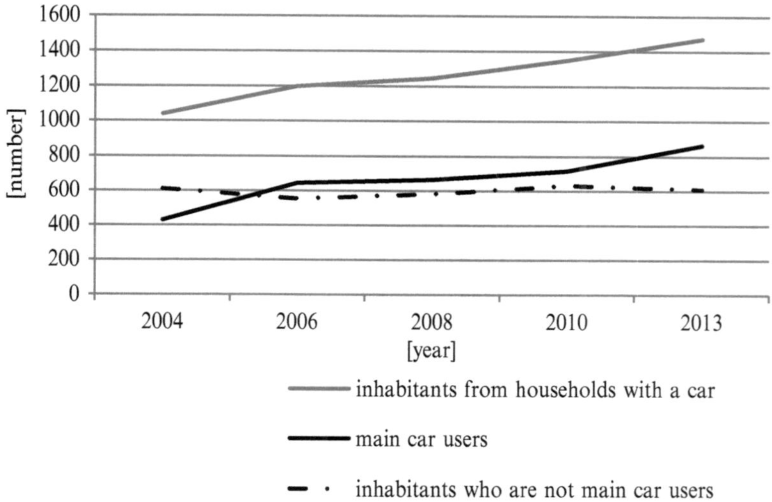

Fig. 4 Motorisation status of the respondents of market research conducted in Gdynia in the years 2004–2013 [number of people] [7–11]

Table 2 The declared way of urban travel by inhabitants of Gdynia in the years 2004–2013 in the light of market research [% of inhabitants] [7–11]

The mode of urban travel	Year				
	2004	2006	2008	2010	2013
Always by car	10.19	17.6	18.93	20.96	25.00
Usually by car	12.07	16.0	16.31	14.53	16.35
Equally by car and public transport	13.74	11.0	11.06	12.40	11.85
Always by public transport	34.23	28.7	27.62	31.34	28.35
Usually by public transport	27.59	26.5	25.67	20.25	18.15
Other	2.18	0.2	0.41	0.52	0.30

The determination of the way in which the respondents travelled on the day prior to the survey was conducted using the method of "a day in picture". The findings also confirm that the way in which inhabitants of Gdynia travel within the city is undergoing change. The growing tendency to choose to undertake urban travel by car can be observed at every step. Within the analysed period, the share of this mode of transportation in actual trips grew by approximately 15 % (Table 3).

The average time of work commute by car was always a little shorter than that of a commute to an educational establishment, namely, by 1–3 min (Fig. 5). The fact that inhabitants believed a door-to-door commute to be nearly two times shorter than the equivalent journey by public transportation is significant.

Table 3 The actual mode of urban travel of Gdynia inhabitants in the years 2004–2013 on the basis of trips fulfilled by the respondents on the day prior to the research (according to the chosen modes of transportation)[a] [%of trips] [7–11]

Way of travel	Year 2004	2006	2008	2010	2013
Car	38.64	47.25	46.95	48.65	53.10
Bus	36.93	31.76	29.94	28.81	25.04
Trolleybus	17.20	14.82	14.72	14.97	14.00
Fast Urban Railway or train	5.78	5.40	7.20	6.32	6.40
Bicycle	0.57	0.27	0.41	0.35	0.80
Other	0.88	0.50	0.78	0.90	0.66

[a]In the case of an indirect trip, it was classified according to the mode of transportation used during the first leg of this trip

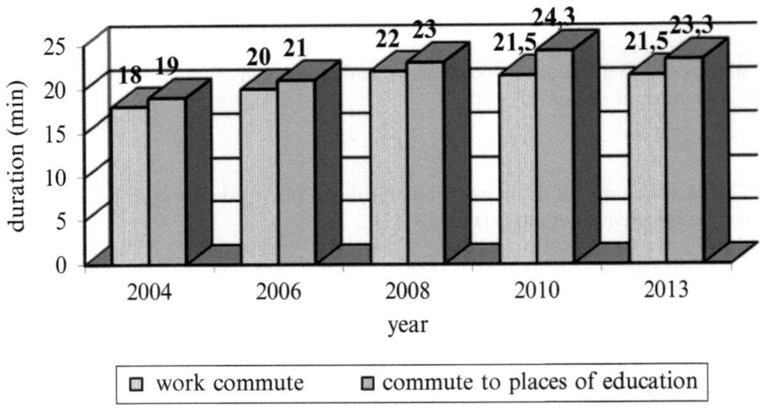

Fig. 5 The average time of car commute to workplace and educational establishments in Gdynia in the years 2004–2013 [min] (7–11)

Determinants of the Choice to Undertake Urban Travel by Car in Gdynia in the Light of Research into Preferences and Travel Behaviour

The factors determining the choice to undertake urban travel by car were always or usually always established during transport behaviour research among the inhabitants of Gdynia. The inhabitants were asked about the reasons why they chose to undertake urban travel by car.

In all years analysed, two decisive factors determining the choice of a car were dominant: shorter trip duration and higher comfort. Generally speaking, regardless of the chosen order of a factor, they constituted as much as 57 % of all answers given by respondents (Fig. 6). It is necessary to point out that the share of inhabitants who highlighted the shorter duration of travel as the reason for choosing to

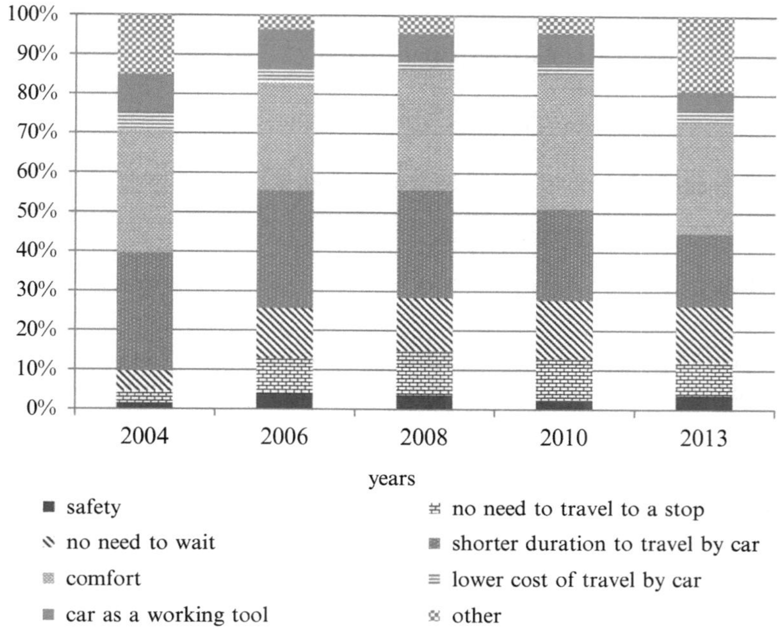

Fig. 6 Structure of the factors determining the choice to undertake urban travel by car in the light of research conducted in Gdynia in the years 2004–2013 [% answers] (7–11)

undertake urban travel by car can be seen to have decreased systematically, starting at 30 % in 2004 and finishing at 17 % in 2013. However, the number of respondents highlighting comfort as the main reason for travelling by car has grown proportionally. The importance of the third factor, namely, the lack of waiting time, also increased from 5 % in 2004 to 15 % in 2010 (13 % in 2013).

It is also worth noting that the respondents expressed no strong feelings about the low cost of car travel and travel by public transportation. The highest number of inhabitants that highlighted this determinant was noted in 2004 and amounted to 5 %. The lowest was in 2008 amounting to 2 %.

The ability to travel directly to one's destination without multiple connections as well as the fact that many respondents carry items such as shopping (9 % and 8 %, respectively), among other factors, can be seen to have grown in significance. This led to the necessity to distinguish such determinants in research conducted in 2013 (Fig. 6).

The order of importance of the reasons for choosing to undertake urban travel by car has undergone changes during the course of research. The following may be counted among the main reasons for choosing to undertake urban travel by car, based on the findings of research from all previously named years, regardless of the order in which they were selected:

- Comfort
- Shorter duration of travel by car
- No need to wait
- Direct travel (no need for multiple connections)

- Carrying items (shopping)
- No need to reach a public transportation stop
- Car as a working tool
- Safety—exposure to violence, i.e. from other passengers
- Lower cost to travel by car
- Other (personal reasons named by the respondents)

Summary

The establishment of the role of cars in urban travel is carried out as part of complex traffic research and research into transport preferences and behaviour of urban inhabitants. As such research is highly complicated, difficult to carry out and costly, it is only undertaken in select cities. By taking advantage of the fact that the findings are stored on the European Platform on Mobility Management, it is possible to compare the role of cars among various European cities. On the basis of this research, it can be claimed that despite the large dependency of the mode of urban travel on local conditions, the growing tendency to undertake urban travel by car is visible in European cities, including Polish cities.

Over 50 % of urban trips in European cities in the past 10 years were undertaken by car. Lone drivers made up nearly 100 % of those travelling during such trips. The largest share of cars was seen in the cities of The United Kingdom and Ireland.

In the Polish city of Gdynia, where research into transport behaviour and preferences was conducted five times in the period between 2004 and 2013, it was established that ¾ of households are currently households with at least one car. Both the declared and actual mode of undertaking urban travel is by car despite the fact that due to increasing congestion, the duration of travel increased by approximately 15 %. Comfort and the shorter duration of travel constitute the main factors determining the choice to undertake urban travel by car. There is no evidence to suggest that in the nearest future, the role of cars in urban travel shall diminish; they have, in fact, become a commonly used mode of transportation for such travel.

References

1. Budziszewski T., Mendel T., Wyniki kompleksowych badań ruchu Gdańsk 2009 oraz przykład ich wykorzystania do szacowania funkcji oporu przestrzeni, In: Kompleksowe badania ruchu. Teoria i praktyka. Doświadczenia polskich miast, Zeszyty Naukowo-Techniczne Stowarzyszenia Inżynierów i Techników Komunikacji Rzeczpospolitej Polskiej Oddział w Krakowie, Seria: Materiały Konferencyjne, vol. 93 (zeszyt 152), pp. 32–33, Kraków (2010).
2. Szarata A., Wybrane wyniki kompleksowych badań ruchu w województwie małopolskim, In: Transport Miejski i Regionalny, vol. 1, pp. 30–34, Stowarzyszenie Inżynierów i Techników Komunikacji Rzeczpospolitej, Kraków (2014).
3. The European Platform On Mobility Management http://www.epomm.eu.
4. Piechociński J., Motywacje podróży i podział zadań przewozowych, www.siskom.waw.pl/komunikacja/motywacje-podrozy.htm.

5. Beim M., Wrocławskie badania ruchu, http://wroclaw.pl/wroclawski_badania_ruchu,1.dhtml.
6. Knapik B., Badania ruchu: coraz więcej aut i rowerów, coraz mniej podróży komunikacją, http://www.tuwroclaw.com/wiadomosci,badania-ruchu-coraz-wiecej-aut-i-rowerow-coraz-mniej-podrozy-komunikacja,wia5-3266-7856.html.
7. Badania pierwotne pt. "Preferencje i zachowania transportowe mieszkańców Gdyni", Katedra Rynku Transportowego Uniwersytetu Gdańskiego i Zarząd Komunikacji Miejskiej w Gdyni, Gdynia (2004).
8. Badania pierwotne pt. "Preferencje i zachowania transportowe mieszkańców Gdyni", Katedra Rynku Transportowego Uniwersytetu Gdańskiego i Zarząd Komunikacji Miejskiej w Gdyni, Gdynia (2006).
9. Badania pierwotne pt. "Preferencje i zachowania transportowe mieszkańców Gdyni", Katedra Rynku Transportowego Uniwersytetu Gdańskiego i Zarząd Komunikacji Miejskiej w Gdyni, Gdynia (2008).
10. Badania pierwotne pt. "Preferencje i zachowania transportowe mieszkańców Gdyni", Katedra Rynku Transportowego Uniwersytetu Gdańskiego i Zarząd Komunikacji Miejskiej w Gdyni, Gdynia (2010).
11. Badania pierwotne pt. "Preferencje i zachowania transportowe mieszkańców Gdyni", Katedra Rynku Transportowego Uniwersytetu Gdańskiego i Zarząd Komunikacji Miejskiej w Gdyni, Gdynia (2013).

Comparison of Road Safety Policy Objectives in Poland and in the European Union

Monika Paradowska

Abstract The aim of the chapter is to identify main actions undertaken within the Polish road safety policy and to assess its efficacy based on the system approach, as well as to present recommendations for enhancing road safety in Poland. In the first part of the chapter key objectives of the EU road safety policy are elaborated. Then, the situation in Poland related to the number of road fatalities and to the causes of accidents is described, and policy instruments in Poland are presented. In the last part of the text an attempt is made to evaluate the Polish road safety policy. Besides, recommendations for improving road safety performance in Poland are elaborated on. In conclusion, it is stated that—according to the system approach—the efficacy of road safety instruments is determined by many factors and requires an integrated set of different actions.

Keywords Road safety in Poland • Road safety policy in the EU

Introduction

Road accidents are considered one of the most important social problems. According to WHO, 1.24 million people die each year in road traffic collisions and more than 50 % of victims are non-motorised transport users. Moreover, injuries resulting from road accidents are among the three leading causes of death for people in the age group 5–44 as well as an important cause of disability worldwide [1]. In fact, direct and indirect economic and social costs of road accidents are definitely more far-reaching. The direct costs are reflected inter alia in expenditures on medical services such as healthcare in hospitals, emergency care and transportation, and other services such as police, fire service, insurance administration, property damage, costs of funerals or judicial proceedings [2, 3]. Examples of indirect costs, sometimes acutely difficult to measure, are production losses due to the lost

M. Paradowska (✉)
Faculty of Economics, Opole University, Opole, Poland
e-mail: mparadowska@uni.opole.pl

M. Bąk (ed.), *Transport Development Challenges in the Twenty-First Century*,
Springer Proceedings in Business and Economics,
DOI 10.1007/978-3-319-26848-4_11

working days, early retirement or fatalities, and the human cost or human value resulting from the loss of quality of life experienced by injured and disabled people [4] as well as moral damages, pain and suffering experienced by family members and friends [5]. The global economic consequences of motor vehicle accidents were estimated at over 500 billion USD—from 1 % up to 3 % of the respective GNP of a country [1].

The scale of the problem of road accidents across Europe has disposed the European Union to undertake actions aimed at lowering the number of road fatalities and at reducing their socio-economic costs. Although decades of more and more conscientious activities on the European and national levels have brought noticeable results, in many countries accidents still arise as a very significant problem. This is the case also in Poland, where in 2013 direct and indirect socio-economic consequences of road accidents accounted for about 2.1 % of GDP[1] [6]. Many tools and actions have been applied in Poland as a response to the key objectives of the European road safety policy. As a result, the total number of road fatalities has declined, however, according to the road safety statistics, Poland is still at the far end of the list of all member states.

As practice shows, a system approach is needed in order to build and implement a coherent and successful strategy for improved road safety. According to the principle of subsidiarity, road safety problems should be solved on the European level only in this case, when it is necessary for efficient solutions on the national level (or when implementation of such solutions on the national level would not lead to expected results). For this reason, many different actions need to be prepared, organised and undertaken in Poland. A system approach consists in clear, interdisciplinary procedures aimed at solving road safety problems, including for example institutional and organisational structures, efficient road safety management system, legal, organisational and financial rules, principles of institutional and organisational cooperation etc. It can be defined as a decision-making process conducted from general to detailed solutions [7].

The EU Strategic Road Safety Objectives at the Beginning of the Twenty-First Century

The last decade of the twentieth century showed significant efforts of the EU to develop common approach towards preventing road accidents and reducing their negative impact on the European society and economy (see for example [8–11]). Although many actions were undertaken, traffic volume has been systematically increasing, and in 2000 there were over 40,000 people killed and more than 1.7 million people injured in road accidents in the EU member states [12]. According to estimations, both direct and indirect costs of accidents accounted for about 160

[1] Costs of both road accidents and collisions accounted in 2013 for 2.9 % of Polish GDP [6].

billion euros, equivalent to 2 % of the EU's GNP [12]. With that in mind and with the vision of the EU's enlargement, new steps were taken within the European road safety policy.

The White Paper "European transport policy for 2010: time to decide" published in 2011 [12] outlined very clearly the goal of halving the number of deaths on the roads till 2010 in comparison to 2001. A new road safety action programme was developed for the years 2002–2010 [13], stressing the need for political commitment for improving road safety across Europe. Each member state was supposed to take individual responsibility in this field, by implementing adequate policy measures and striving for the best results. Main directions of actions at the European level assumed harmonisation of penalties, road signs and blood-alcohol levels as well as further development of new technologies, e.g. in terms of ensuring safer transport systems for pedestrians and cyclists. Seven core areas of actions were outlined in the 3rd European action programme for road safety (RSAP). Focus on monitoring and reporting resulted inter alia in the development of institutional foundations in the form of creation of the Road Safety Observatory (ERSO) with the mission to collect, analyse and disseminate road safety data and knowledge [14], as well as in setting up the European Road Safety Charter (ERSC) and ETSC's Road Safety Performance Index (PIN) in 2006 [15].

The goal of reducing the number of fatalities by 50 % by 2010 was not fully completed. However, striving for more satisfactory results in terms of road deaths and their socio-economic costs was even strengthened at the beginning of the new decade. In 2010 the 4th RSAP [16] was published entitled "Road safety: Policy orientations on road safety 2011-20",[2] with the sustained goal of halving the number of fatal road accidents in the EU between 2010 and 2020. Similarly as in the previous period, member states are expected to take the key responsibility for achieving this objective, to contribute to a better road safety performance at the European level, and to act while taking three main principles into account: striving for the highest road safety standards, an integrated approach to road safety as well as subsidiarity, proportionality and shared responsibility [16]. Seven strategic objectives were identified by the Commission as a guideline for actions developed on both the EU and national levels: (1) to improve education and training of road users, (2) to increase enforcement of road rules, (3) safer road infrastructure, (4) safer vehicles, (5) to promote the use of modern technology to increase road safety, (6) to improve emergency and post-injuries services and finally (7) to protect vulnerable road users.

The main objective of the European road safety policy is also highlighted in the White Paper "Roadmap to a Single European Transport Area—Towards a competitive and resource efficient transport system" [17], which was published a year after the 4th RSAP. Moreover, the document provides much more ambitious goal in the form of the "zero-vision", i.e. of a nearly total reduction of road fatalities by 2050. For the first time the target of reducing serious traffic injuries was set. Four action

[2] According to ETSC, this document fell short of the expectations and did not provide a clear vision and priorities, which could be the basis for creating appropriate instruments and initiatives [11].

areas are pointed out in the White Paper, including road safety technology, road injuries and emergency services, training and education and particular attention to vulnerable users such as pedestrians, cyclists and motorcyclists [17]. Some of instruments and actions are supposed to contribute to combined goals indicated in the White Paper, including improving road safety, lowering greenhouse gases emission from the transport sector and/or reducing oil dependence.

Due to the undertaken actions, significant progress has been made over recent years in terms of improving road safety in the EU. Although the rate of road accidents deaths was reduced by the total of 43 % between 2001 and 2010, the ambitious goal of 50 % reduction was achieved in 2012 [17, 18]. New legislation was developed and introduced,[3] new initiatives across Europe were taken in order to enhance positive changes in drivers' behaviour as well as to develop and disseminate most successful practices. Focus has been also on research and development, especially in terms of the development of new technology supporting vehicle safety, avoidance of road accidents and speeding up emergency response times. Financial support and improved monitoring and reporting have been of great importance for meeting national and regional objectives (see for example [19, 21, 22]).

In 2000–2010 the volume of road passenger traffic increased by more than 5.2 % and the volume of passenger car traffic by more than 8.3 % (in 2000–2012 the increase in volume accounted for over 5.8 % and 3.7 % respectively). More dynamic increase could be observed in the volume of road freight traffic: in the period 2000–2010 it accounted for nearly 16 % and in the years 2000–2012 for more than 11.2 % [23]. However, consistent striving for better road safety resulted in satisfactory effects, even if achieving the target of halving the number of road deaths was slightly delayed. Figure 1 presents the number of road fatalities in the EU28, the EU15 and

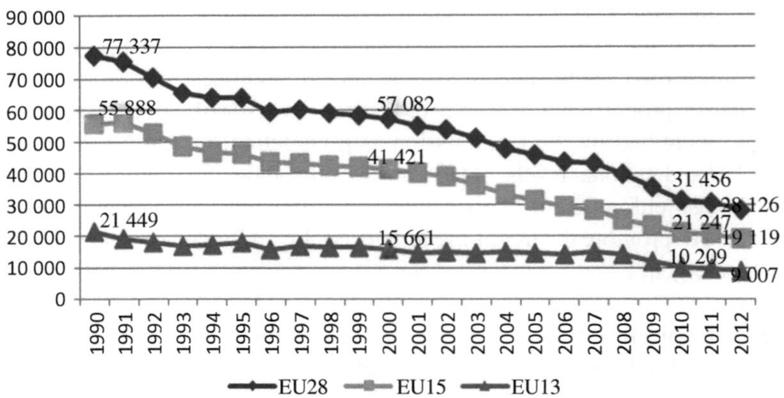

Fig. 1 The number of road fatalities in the EU between 1990 and 2012 (Source: own elaboration based on EU transport in figures. Statistical pocketbook 2014. European Union, Belgium (2014), http://ec.europa.eu/transport/facts-fundings/statistics/pocketbook-2014_en.htm)

[3] A general overview of the EU's legislation related to road safety is available at [20].

in the EU13 between 1990 and 2012. What is worthy of mentioning is that the most visible effects were worked out in the EU15—a 53.8 % reduction in the number of road fatalities between 2000 and 2012. In contrast, the poorest road safety performance can be observed in the new member states, where the total reduction in the number of road fatalities accounted for 34.8 % in the period 2000–2010 and for 42.5 % in the period 2000–2012 [23].

Similarly, in the current decade, regional differences are visible in terms of the road safety performance and of achieving the "second" European road safety target of halving the number of road fatalities. In 2013 slightly more than 26,000 people died in the EU in road accidents, which marked a decrease of around 8 % in comparison to 2012. However, the safest roads were characteristic for the UK, Sweden and Denmark and the least satisfactory statistics on road accidents were found in Romania, Latvia, Poland and Lithuania. On average, the number of road deaths was reduced by 18 % in the EU28 in 2010–2013, but in Poland the reduction accounted for slightly less than 15 % [24].

The year 2014 turned out to be disappointing when analysing progress made on the way to safer roads in the EU. There were no changes in the average reduction of the number of road fatalities—it remained on the level of 18 % in comparison to 2010 and accounted for 1 % in comparison to 2013 [25]. However, the European Commission continues to persist and has been continuingly working on improving, developing and implementing new solutions on the European level towards better road safety. But, according to the principle of subsidiarity, much has to be done on national and regional levels. And in some member states, including Poland, the progress seems to be unsatisfactory.

Road Accidents in Poland: The Unsolved Problem

Polish Road Safety Statistics

Poland still belongs to the member states with the worst road accidents performance. The goal of halving the number of road fatalities has not been achieved so far—in 2010 there was a nearly 38 % reduction of road deaths in comparison to 2000 and in 2013 a 40 % reduction in comparison to 2003 (due to the accession of Poland to the EU in 2004, the target of halving traffic deaths has been set for 2013 in comparison to 2003). A positive effect is that the number of accidents and road fatalities in Poland has been systematically decreasing—in 2000 there were 6294 and in 2013—3351 road deaths reported. This downward trend has been continued in spite of the fact that the motorisation rate and the demand for road traffic has simultaneously increased [23].

Table 1 shows data related to the number of fatalities per million inhabitants, per 10 billion passenger-kilometres and per million passenger cars in 2007–2012 in Poland and in the EU in comparison to the "best" and to the "worst" countries. Poland moved down to lower positions in all three rankings, even though the country reduced the number of road fatalities by 37 % in this period. This situation can

Table 1 Selected statistics on road safety in Poland (2007–2012)

2012			2011			2010			2007		
Region / country	Fatalities per million inhabitants	position (total = 29)	Region / country	Fatalities per million inhabitants	position (total = 28)	Region / country	Fatalities per million inhabitants	position (total = 28)	Region / country	Fatalities per million inhabitants	position (total = 28)
EU28	56	11	EU27	60	11	EU27	62	11	EU27	86	12
PL	93	27	PL	109	29	PL	102	25	PL	146	26
MT	26	1	UK	31	1	SE	28	1	MT	29	1
RO	102	29	PL	109	28	EL	111	28	LT	219	28
Region / country	Fatalities per 10 billion pkm	position (total = 29)	Region / country	Fatalities per 10 billion pkm	position (total = 28)	Region / country	Fatalities per 10 billion pkm	position (total = 28)	Region / country	Fatalities per 10 billion pkm	position (total = 28)
EU28	59	14	EU27	61	13	EU27	64	12	EU27	88	12
PL	169	28	PL	131	25	PL	129	24	PL	229	23
SE	26	1	SE	29	1	SE	27	1	UK	44	1
RO	254	29	RO	259	29	RO	303	28	RO	448	28
Region / country	Fatalities per million passenger cars	position (total = 29)	Region / country	Fatalities per million passenger cars	position (total = 28)	Region / country	Fatalities per million passenger cars	position (total = 28)	Region / country	Fatalities per million passenger cars	position (total = 28)
EU28	115	14	EU27	126	13	EU27	131	16	EU27	187	13
PL	194	23	PL	237	25	PL	232	23	PL	399	22
MT	44	1	UK	67	1	SE	62	1	MT	54	1
RO	463	29	RO	466	28	RO	555	28	RO	782	28

Source: EU transport in figures. Statistical pocketbooks 2009–2014. European Union, Belgium (2009–2014), http://ec.europa.eu/transport/facts-fundings/statistics/

be considered as a sign of insufficient efforts and inefficient Polish policy towards improving road safety in comparison to the majority of other EU member states.

Statistically, in Poland every fourth death results from road accidents. Moreover, road accidents are the main cause of the death of males under 44 years of age [26]. What is also important is that the majority of road accidents take place during the day and in good weather conditions. Much more road accidents occur in the built-up areas than in the non-built-up areas (72.7 % and 27.3 % respectively in 2013), though more people are killed in road accidents in the non-built-up areas (52.9 % in

2013 and 54.2 % in 2014). In 2014 the largest groups of victims were people over 60 years of age (28.6 %), people aged 40–59 (27.8 %) and 25–39 (23 %). But still the problem is that a lot of young people die due to road fatalities—553 (16.5 %) in 2013 and 499 (15.6 %) in 2014 [27, 28].

Causes for Accidents in Poland

The definite majority of road accidents in Poland are caused by passenger car drivers (63 % in 2014), leading to nearly a half of road accident deaths (48.7 % in 2014) and the majority of injuries (67.6 % in 2014). Drivers of goods vehicles caused in 2014 5.2 % of all accidents and were responsible for 6.6 % of road fatalities and 5.5 % of injuries. Road accidents caused by pedal cyclists and motorcyclists accounted respectively for 5.1 % and 2.9 % of all road accidents, in which 297 (9.3 %) people were killed and 2777 (6.5 %) injured. A special attention should be paid to the group of young drivers (aged 18–24) who caused 17 %% of all road accidents in 2014, including nearly 16 % of road fatalities and 19.5 % of injuries. Pedestrians were responsible for 8.7 % of all accidents in 2014 and for 17.6 % of all road traffic deaths [28].

When analysing different types of road accidents in Poland in 2014 and the number of people killed (see Fig. 2), it turns out that the main reason for accidents are car crashes, including side, head-on and rear-end collisions. 1130 people died in car crashes in 2014, what accounted for 39.9 % of the total number of road fatalities. The second most significant reason for road deaths is the incorrect behaviour of drivers towards pedestrians and incorrect behaviour of pedestrians (34.5 % of all road deaths in 2014) [28].

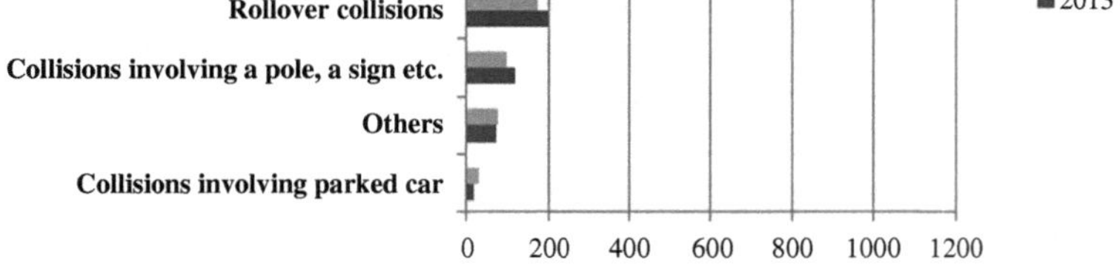

Fig. 2 The number of road accident deaths in Poland in 2013 and 2014 according to the type of road crashes. (Source: Symon, E.: Stan bezpieczeństwa ruchu drogowego w Polsce w 2014 roku (Road safety performance in Poland in 2014). Komenda Główna Policji, Biuro Ruchu Drogowego, Zespół Profilaktyki i Analiz, Warsaw (2015), p. 16)

As indicated above, the biggest share of road accidents in Poland is caused by drivers. And year by year there are the same main causes for accidents, comprising: inappropriate speed for road conditions, incorrect overtaking, non-compliance with vehicle right of way and incorrect behaviour towards the pedestrian, accounting respectively for 40 %, 16 %, 15.1 % and 12 % of accidents caused by drivers in 2014 [28]. The same categories of causes prevail (with similar shares) in terms of road accidents caused by motorcyclists. The behaviour of pedal cyclists leads to accidents and road deaths mostly due to non-compliance with vehicle right of way, incorrect turning, speeding, unsafe lane changing and incorrect cycling through pedestrian crossings. Incorrect behaviour of pedestrians reveals itself mostly in unsafe entering or crossing the road, crossing the road in areas where pedestrian access is not allowed and ignoring the red light signal [28].

The number of road fatalities caused by drunk drivers is still alarming. In total 2579 (7.3 %) road accidents and 363 (11.3 %) traffic deaths were caused in 2014 by traffic participants under the influence of alcohol, including 1838 (5.2 %) accidents and 256 (8 %) road deaths when a drunk driver was the culprit.

The level of the road safety depends also on driving culture. There would be for sure less road accidents in Poland and the number of road deaths would be lower if the Polish drivers were less aggressive, less distracted and generally if they followed traffic regulations more accurately.

Statistical data related to road accidents are of great importance, though they cannot provide the full image and background of insufficient road safety in Poland. The Supreme Audit Office in Poland (Najwyższa Izba Kontroli, NIK) has conducted a detailed research on the causes for poor road safety in 2000–2010. According to the research results, the most significant factor endangering the lives of people on Polish roads is the poor performance of infrastructure [29]. In many European countries there is a traffic ban on roads that are in disastrous condition unless the infrastructure is renovated and modernised. However, in Poland such a movement would lead to the traffic ban on nearly half of the roads [30]. Moreover, although the motorway network density in Poland increased by over 100 % in 2007–2012, it is still far from the EU average [31]. Incorrect driver education and trainings are in the second position among main reasons for inefficient road safety. Drivers with little experience (soon after passing the driving exam) account for about 30 % of all culprits of accidents. Old cars and vehicles in generally bad condition are frequently used in Poland, which also contributes to the poor road safety. What is a little bit surprising is that bad and/or unsatisfactory traffic organisation was also listed in the report, though there had been several years of NIK's controls and recommendations for improvements [29]. Finally, there were insufficient monitoring and conditions for transport of hazardous materials, as well as poor safety at railroad crossings resulting from insufficient cooperation between road and rail managers indicated in the NIK's report. Another report on the progress made towards better road safety in Poland was published in 2013 [32], and it is discussed more widely in Sect. 4.

The document prepared by the Supreme Audit Office shows a detailed list of arguments that not only Polish drivers and other traffic participants are responsible for the poor road safety in Poland. The Polish road safety programmes elaborated in the next section have been adjusted to the main road safety policy objectives of the EU. However, this is a necessary but insufficient condition for improving road safety in Poland. As mentioned in the introduction, Poland needs also a system approach for establishing efficient road safety management and control system which would ensure meeting the goals set out in the road safety programmes (see Sect. 3.3). Better and more reliable institutions, greater political and institutional commitment on national, regional and local levels as well as investments in infrastructure, including infrastructure measures for traffic safety, are basically needed for making progress.[4]

Road Safety Policy in Poland

The beginning of the Polish road safety policy dates back to the early 1970s, when the first Polish legal act on measures towards better road safety was adopted [33]. However, the policy had not been very strict for the next several years. For this reason, many recommendations were suggested in the report prepared by the World Bank in 1992 [34]. This led to the creation of the National Road Safety Council (Krajowa Rada Bezpieczeństwa Ruchu Drogowego, KRBRD) in 2002 [35]. Three programmes were launched between 1995 and 2005 in order to prevent road accidents and improve road safety in Poland: GAMBIT 96, GAMBIT 2000 and GAMBIT 2005.

GAMBIT 96—"Integrated Programme for Improving Road Safety"—was launched in 1996 and lasted 3 years. Its evaluation in 2000 highlighted the need for a verified approach to the road safety actions in Poland. This was reflected in the next edition of the project entitled GAMBIT 2000. It outlined the key policy measures for the years 2001–2010. The strategic objective of the programme assumed the reduction of the number of people killed on Polish roads to 4000 in 2010. It focused on seven problem areas, setting foundations for the long-term road safety policy and building public acceptance for improving road safety [36]. Accession of Poland to the EU in 2004 changed Polish road safety goals and was the main cause of launching the new programme—"National Programme for Improving Road Safety 2005-2007-2013 GAMBIT 2005" [37]. The strategic target of halving the number of road deaths was adapted as one of the Polish road safety objectives. However, the timeframes were adjusted to the specific conditions of the new member state and the deadline for achieving the target of reducing the number of people killed on roads by a half was set for 2013 as compared to 2003. Thus, no more than 2800 of road deaths should be reported in 2013 in Poland (this goal was achieved only in 83 %). There was a set of road safety measures specified in the document with five detailed goals presented in Fig. 3 [37]. One year after the accession of Poland to the EU the Polish Ministry Council approved a new project of transport

[4] The issues of main problems leading to and resulting from the poor road safety in Poland are discussed in more detail in Sect. 4.

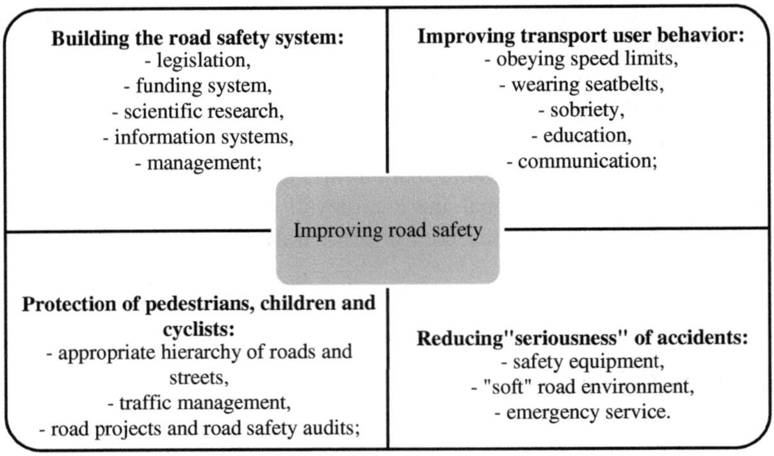

Fig. 3 Elements of system approach according to the Polish road safety policy objectives outlined in GAMBIT 2005. (Source: own elaboration based on Stomma, P.: Odpowiedź podsekretarza stanu w Ministerstwie Transportu—z upoważnienia ministra—na interpelację nr 2287 w sprawie planów rządu mających na celu poprawę stanu bezpieczeństwa ruchu drogowego (The answer of the vice secretary of the state in the Ministry of Transport—on the behalf of the minister—to the interpellation 2287 related to the government plans aimed at improving road safety). Warsaw (2006), http://orka2.sejm.gov.pl/IZ5.nsf/main/74EC5E95)

policy entitled "National Transport Policy for 2006-2025". Improving transport safety, especially in road transport in the form of "radical reduction of road fatalities", was included into four key objectives [38].

Actions taken under GAMBIT 2005 can be considered as an attempt to build a system approach towards accidents prevention and mitigation of their negative consequences [39]. According to the system approach, detailed actions and measures in different but interrelated areas should have contributed to the achievement of general assumptions. Moreover, the central programme has been followed by few programmes on the voivodeship and several programmes on the county levels.

The system approach was a basic guideline followed while preparing the new National Road Safety Programme 2013–2020. Six principles were formulated as necessary for successful implementation of the programme, including "4xE" Approach [(1) Engineering, (2) Enforcement, (3) Education, (4) Emergency] [40]. With the Vision Zero as the key long-term target, two main objectives were highlighted in the document: to limit the annual number of fatalities by at least 50 % until 2020 (the maximum of 2000 of fatalities in 2020), as well as to limit the annual number of seriously injured by at least 40 % until 2020 (no more than 6900 of seriously injured in 2020). Main pillars of the Polish road safety programme were based on the EU's (see for example [16, 17]) and the UN's (see for example [1]) road safety policies, as well as on the National Development Programme until 2020 [41] where road safety was indicated as one of the most deserved areas of action in terms of enhancing safety of Polish citizens. Five pillars listed in the National Road Safety Programme 2013–2020 (see Fig. 4) are expected to be the base for more detailed

Safe Road User	Shaping safe behaviours of road traffic users
	Protection of road users
Safe Road	The implementation of the road safety standards eliminating the most serious hazards in road traffic
	Development of the road infrastructure safety management system
Safe Speed	Shaping drivers' behaviours in relation to driving at a safe speed
	Making the speed management system more efficient
Safe Vehicles	Enhancement of actions regarding vehicle technical condition inspections
	Improvement of safety systems in vehicles
Rescue service & post-crash response	Integration and Development of National Rescue System
	Transformation of the support of victims of road accidents system

Fig. 4 Priorities and directions of actions within the five pillars listed in the National Road Safety Programme 2013–2020. (Source: own elaboration based on Narodowy Program Bezpieczeństwa Ruchu Drogowego 2013–2020 (National Road Safety Programme 2013–2020). Krajowa rada Bezpieczeństwa Ruchu Drogowego (2013), pp. 23–51, http://www.krbrd.gov.pl/files/file/Programy/KRBRD-Program-P1a-20140422-S2-K3d-EN.pdf)

actions in three priority directions reflecting fundamental problems of road safety in Poland—engineering, enforcement and education. According to the authors, the National Road Safety Programme 2013–2020 should be considered a set of policy directions rather than a detailed action plan. In order to strive for a general vision and to implement assumptions of the programme, more detailed action plans were proposed for the year 2013 and 2014–2015 [42, 43] with agenda, schedule and ranking of actions, and indicators necessary for appropriate monitoring.

Several years of more advanced preparation and more restricted implementation of road safety programmes in Poland, which was frequently shaped by the European and international policy trends and recommendations, contributed to the evolution of the organisational structure of the appropriate institutions (see Table 2).

The key institution is the National Road Safety Council working as an auxiliary inter-department body of the Polish Council of Ministers for road safety issues [35]. Counterparts of the council on the regional level are 16 Provincial Road Safety Councils (Wojewódzka Rada Bezpieczeństwa Ruchu Drogowego, WRBRD). The Road Traffic Safety Centre (Centrum Bezpieczeństwa Ruchu Drogowego) at the Motor Transport Institute (Instytut Transportu Samochodowego) plays also an important role. Its main tasks embrace research, analyses and collecting statistical data as well as education, consulting and evaluation of selected road safety measures. The RTS Centre cooperates closely with the Polish Parliament and the government institutions, local governments and non-governmental organisations [44]. A very significant initiative of the Motor Transport Institute was the creation of the Polish Road Safety Observatory (Polskie Obserwatorium Bezpieczeństwa Ruchu

Table 2 Key road safety actors per function in Poland

Key functions	Key actors
1. – Formulation of national RS strategy – Setting targets – Development of the RS programme	– The Ministry of Transport: responsible for road safety and supervises the DG of National Roads and Motorways and the Chief Inspector of Road Transport – The National Road Safety Council (NRSC): inter-ministerial body and lead agency for road safety – Regional Road Safety Councils (in 16 regions): development of regional road safety programmes.
2. Monitoring of the RS development in the country	– NRSC – Motor Transport Institute – The National Police Headquarters
3. Improvements in road infrastructure	– Ministry of Transport (national roads) – Ministry of Administration and regional authorities (regional roads) – Local authorities (lower class of roads) – City authorities (urban roads)
4. Vehicle improvement	– Ministry of Economy, Ministry of Transport – Transport Technical Supervision – Polish Chamber of Vehicle Check-up Stations – Motor Transport Institute
5. Improvement in road user education	– Ministry of Transport (driver education) – Polish Federation of Driving Schools Associations – Ministry of Education (children education) – Motor Transport Institute
6. Publicity campaigns	– NRSC – DG of National Roads and Motorways – Police – Local authorities and other stakeholders – NGO
7. Enforcement of road traffic laws	– Regional Chiefs of the police: main responsibility for road safety – The National Police Headquarters: launch of national schemes, co-ordination of international actions.
8. Other relevant actors	– Regional and local road authorities – Regional Road Safety Centres: drivers' examination centres and other road safety activities – Private NGOs, e.g. Motor Polish Association, Automobile Clubs, Road and Safety Association, Associations of Support for Road Accidents Victims – Research organisations: the Motor Transport Institute, Gdansk University of Technology, Krakow University of Technology, Warsaw University of Technology, Road and Bridge Research Institute, the Institute of Forensic Research from Krakow.

Source: Road safety country profiles. DG-TREN, European Commission (2005), http://ec.europa.eu/transport/road_safety/observatory/country_profiles_en.htm; Technical Assistance in support of the Preparation of the European Road Safety Action Program 2011–2020. Final Report. DG-TREN, European Commission, Brussels (2010), http://ec.europa.eu/transport/road_safety/observatory/doc/drl_rules.pdf, cited by: Road Safety Management Profile. Poland. DaCoTa, Project co-financed by the European Commission, Directorate-General for Mobility and Transport (2010), http://ec.europa.eu/transport/road_safety/specialist/knowledge/dacota/pdf/rsm/dacota-rsm-profile-template-po_en.pdf

Drogowego) in 2010–2013, which was based on the European Road Safety Observatory [45]. Many tasks and responsibilities related to road safety in Poland are dispersed among police, self-governments, institutions focused inter alia on issues related to road management and maintenance, such as for example General Director for National Roads and Motorways (Genralna Dyrekcja Dróg Krajowych i Autostrad, GDDKiA) and its regional chapters or local road management agencies, or on issues related to drivers' education etc. There are also many other institutions and organisations which are indirectly decisive or responsible for road safety, and yet their impact on building a system approach is crucial (see Table 2).

Evaluation of the Polish Road Safety Policy

Both implementation and effects of the Polish road safety programmes, and the organisational structure of institutions and agencies involved in improving road safety have been evaluated in various analyses and reports prepared by different stakeholders on national and international levels.

On the national level, the results of the consecutive road safety programmes have been evaluated mostly by scientists, researchers, and experts representing different institutions responsible for, or interested in, improving road safety in Poland. There is no clear evaluation of GAMBIT programmes available in the form of an official document. However, increasing public awareness and pressure on developing Polish road safety policy, mostly in accordance with the EU's objectives, resulted in many scientific papers devoted inter alia to the assessment of effects of road safety measures as well as in the report of the Supreme Audit Office discussed briefly in Sect. 3.2. When summarising the most significant effects of the GAMBIT programmes, several achievements need to be pointed out [46]:

- Road safety councils were created on the national and regional levels.
- Road safety programmes on national and regional levels are prepared and being implemented.
- More and more specialised and educated scientists and experts are dealing with issues related to improving road safety, at universities, research entities, self-government entities, etc.
- The police database on road accidents has been created and the system of data collection and dissemination is being developed.
- Many institutions and agencies represent willpower and determination to continuously take actions towards improved road safety in Poland.

On the other hand, various weaknesses can be distinguished, hampering successful implementation of road safety measures and meeting the targets established in the road safety programmes [46]:

- There is still insufficient number of inspectors and specialists on road safety in central government and self-government institutions on national, regional and local levels.

- Appropriate legal regulations are missing, especially in terms of creating efficient organisational structures for improving road safety, as well as in terms of drivers', pedal cyclists' and motorcyclists' trainings.
- The development of local databases is necessary for better analyses and monitoring of the road safety.
- Support of mass media is not efficient in order to enhance appropriate change of road transport users.
- Wider scientific and research support should be provided for improving road safety.

Although the goal of halving the number of road traffic deaths in 2013 as comparison with 2003 was clearly indicated in the programme GAMBIT 2005, it has not been achieved yet. There were different actions taken of legislative, educational, infrastructural and other character. Only 84 from among 144 tasks (58 %) have been realised. Moreover, some of them turned out to be unsuccessful or inappropriately organised. However, according to estimations, around 6000 people have been saved from traffic death and financial savings have accounted for ca. 34.5 billion PLN within the 7 years of the realisation of the programme [49]. Table 3 presents the overview of positive effects of GAMBIT 2005 as well as of failures and actions which were not accomplished.

Even though the system approach was applied during the preparation of GAMBIT 2005 and different necessary actions were in the document, the measures implemented and goals realised can be considered as partial and insufficient. This is the main reason for not meeting the strategic objective of the Polish road safety policy in 2013. First of all, efficient institutional conditions for the realisation of the strategy have not been provided, and the lack of a coherent and efficient system of financing led to the partial realisation of the programme. Moreover, the implementation of measures has not been correctly monitored and evaluated. To put it simply, implementation and realisation of GAMBIT 2005 were not supported to the extent enabling its effects to be sufficient and satisfactory. In 2013 another report on road safety was published by the Supreme Audit Office [47]. As the document states the development and modernisation of the road infrastructure, the development of the traffic management, the improved emergency services and some selected actions in the road safety management are the key positive achievements in terms of building a consistent road safety system in Poland. However, many other aspects were insufficiently developed or neglected, including technical performance and signposting, the functioning of road traffic supervision, as well as the lack of a common, coherent road safety system. Similar but much more detailed conclusions and arguments were listed in the report prepared by the World Bank in 2013 [48]. The report led to the preparation of recommendations for the transition of the Polish road safety system (see Sect. 5).

However, the lessons learnt shall be enough for conclusions, elimination of failures and for implementation of these actions and measures which determine success of the Polish road safety policy. The new National Road Safety Programme established analogously ambitious target of halving the number of traffic deaths. Meeting

Table 3 Brief evaluation of the effects of the programme GAMBIT 2005

Positive effects	• More training hours for candidates for drivers and couches, more appropriate trainings and exams • Higher effectiveness of traffic supervision and control, police has been supported by the Road Transport Inspectorate and municipal guards since 2002 • More regular and systematic activities of the police • New techniques of traffic control, including technical control of vehicles and more efficient control of behaviour of traffic participants (e.g. in terms of obeying speed limits or working time of professional drivers) • Regulation of the allowed behaviour of pedal cyclists, more efficient protection of pedal cyclists • Lower speed (50 km/h) allowed in built-up areas (2004) • Social campaigns in mass media, e.g. implementation by the GDDKiA medium-term programmes aimed at elimination of very dangerous roads, campaign conducted by Polish Railways aimed at less dangerous railway crossings • Creation and implementation of road safety programmes on the level of voivodeships and counties • Creation and implementation of road safety programmes for the police and for national roads • Creation of the Polish Road Safety Observatory and of two regional observatories • Construction of highways and motorways, modernisation of roads managed by self-governments (co-financed by the EU), construction of safe crossings and implementation of traffic calming measures • Partial application of road safety audit of projects and programmes • Partial modernisation of emergency services.
Negative effects/ actions not accomplished	• Higher speed allowed on highways and motorways • Improved structures of central institutions, especially of the National Road Safety Council • Implementation of the system of protecting pedestrians • Improved programming of road safety, improved creation of modern databases and road safety monitoring on all administrative levels • Providing legal and financial conditions for creation of the training system for professionals, as well as for creation of local executive institutions • Providing conditions for modernising road traffic management (except of highways and motorways) • Building safe road lanes (road safety zone) • Systematic implementation of obligatory methods of road safety management on roads managed by self-governments • Implementation of the system of financing road safety measures and emergency services.

Source: Jamroz, K., Michalski, L.: Analiza możliwości realizacji celów strategicznych bezpieczeństwa ruchu drogowego w Polsce (The analysis of the possibilities of realisation of the strategic road safety objectives in Poland). Prace Naukowe Politechniki Warszawskiej 96 (2013), preprint; Michalski, L.: Wstępna ewaluacja interim Krajowego Programu Bezpieczeństwa Ruchu Drogowego GAMBIT 2005 (Preliminary evaluation of the National Road Safety Programme GAMBIT 2005). Presentation at the International Seminar on Road Safety GAMBIT 2012, http://www.gambit.fril.org.pl/files/Gambit_2012/C_-_S2_-_1_-_PL_-_L._Michalski_-_Wst%C4%99pna_ewaluacja_interim_Krajowego_Programu_Bezpiecze%C5%84stwa_Ruchu_Drogowego_GAMBIT_2005_.pdf

this target seems to be available, provided that the main obstacles are removed and that the road safety policy is realised in a more efficient way [49].

The poor road safety performance in Poland has been one of reasons for more or less intensive and detailed external evaluations which also give useful guidelines for improvements. According to the assessment of institutional and organisational conditions for successful road safety policy making conducted within the project DaCoTa, the main problems are the form of insufficient political will, inadequate competences of the National Road Safety Council as well as the lack of systematic evaluation. All of these factors has led to insufficient road safety and, in turn, to alarming road safety statistics in Poland [50]. As a result, weak road safety management, poor road safety culture among road participants and poor road standards, thee lack of facilities for pedestrians and of ring roads are still the main causes for road accidents in Poland [51].

Unsatisfactory, or at least inefficient improvements in terms of both road safety management structures, management process and policy making shall serve as guidelines and as a starting point for the system change not only in terms of setting policy objectives but also in the context of providing efficient conditions for organising and conducting road safety actions. Thus, a system approach needs to be continuously implemented in the next stages of the process of improving road safety (planning—organisation—implementation—monitoring). Due to the scale of the problem, the Supreme Audit Office and the World Bank have cooperated closely in order to work out recommendations for consistent and complex transition to substantially safer roads in Poland. This issue is discussed in more detail in the next section.

Recommendations for Enhancing Road Safety in Poland

On behalf of the Polish Ministry of Infrastructure and Development the analysis of the road safety management structures in Poland has been conducted since September 2014. The World Bank has coordinated the analysis in cooperation with the core Polish institutions responsible for road safety and prepared the final report.

According to the results of research and studies conducted so far, numerous actions and changes need to be implemented in order to enhance the whole road safety system and to improve road safety in Poland. Otherwise, it will be very difficult to achieve both the goals established in the National Road Safety Programme 2013–2020 and set out be the EU in the White Paper on transport and in the Policy Orientations on Road Safety 2011–2020.

First of all, the Supreme Audit Office [47] suggested that the efforts undertaken by the state in terms of building conditions for enhanced road safety need to be much more dynamic. For this reason, the primary actions should be focused on building coherent organisational and institutional structures in the form of "the Road Traffic Safety Management and Control System" [47] with one public body responsible for all actions and measures aimed at road safety. The second condition for improving road safety is to ensure stable financing. Thorough reorganisation of the road traffic supervision system is also needed, including for example incorporation of the Road Transport Inspectorate into the police, and limited rights of city and municipal guards in terms of speed control. The Supreme Audit Office highlighted the necessity of creating a system of planning and analysing effects of actions taken, which should comprise for example research and scientific entities, detailed databases (e.g. on causes for accidents and efficiency indicators for road safety measures), as well as one main institution responsible for research, education and trainings of the police and professionals. According to the recommendations, changes in legal rules shall in the first place consist in changes in the procedure of imposing speeding tickets. Further modernisation of infrastructure, improved control of the mechanical state of vehicles, and continued development of emergency services were listed in recommendations as well. In the last position there were highlighted the necessity of education and promotion of the required, correct behaviour of road transport participants, including drivers, pedestrians, pedal and motorcyclists.

From the report of the Supreme Audit Office a vision of a renewed, improved road safety management system in Poland is arising with one leading institution which would successfully conduct and coordinate the road safety policy, supervise the creation and implementation of road safety programmes and actions etc. The need of establishing one leading institution responsible for the Polish road safety policy was underlined as well in preliminary guidelines prepared by the World Bank in December 2014 [52]. The National Road Safety Lead Agency in Poland should ensure appropriate social and political support, as well as funding and resourcing, in order to match the particular Polish road safety circumstances. As the World Bank argues [52], the core functions of the National Lead Agency should embrace results focus, i.e. "pivotal institutional management focus", coordination and legislation, funding and resource allocation, promotion and education, monitoring and evaluation, and finally research, development and knowledge transfer.

According to the World Bank's guidelines, the lead agency should operate as an independent body within the Prime Minister's sphere, and the analytical and reporting function should be directed to the Prime Minister. The current position and potential reorganisation of the National Road Safety Council should be carefully considered and special actions should be taken in order to avoid dual leadership of the National Road Safety Council and the newly established lead agency. The suggested preliminary structure of the National Lead Agency shall be systematically transformed according to the Poland-specific circumstances, its functions and operations.

In conclusion, a number of guidelines and recommendations have been provided, especially by the Polish Supreme Audit Office and by the World Bank. Their implementation seems necessary for a more successful road safety policy. Reorganisation of both the national road safety management and control system seems to be crucial and ineluctable, as previous road safety programmes turned out to be difficult in implementation and were not fully accomplished in terms of the postulated goals. Thus, the National Lead Agency should play the primary role in the context of ensuring institutional foundations and sufficient political will, which in turn should result in coordinated and efficient implementation of the road safety measures.

Conclusion

The poor road safety performance in Poland results from many different factors. Incorrect behaviour of Polish drivers, but also underdeveloped infrastructure, insufficient road safety management structures etc. lead to an unsatisfactory position of Poland among other member states in terms of road safety statistics. The system approach which was the core guideline while preparing GAMBIT 2005 and the National Road Safety Programme 2013–2020 turned out to be insufficient, mostly because of the lack of the system approach when implementing the planned measures. The interrelation between objectives set and the political will to achieve them seemed to be too weak. For this reason, the National Lead Agency is expected to provide sufficient institutional conditions for efficient road safety policy, provided that it is empowered with more competences than the National Road Safety Council. However, reorganisation of the institutional road safety structures in Poland needs to be followed by a strict and consequent realisation of different tasks and actions, which were outlined in the National Road Safety Programme 2013–2020, and should be clearly developed in the next detailed action programmes.

All the changes needed for the transition towards better road safety in Poland are long-lasting. Preparations for establishing the precursory lead agency in Poland began at the beginning of 2015. However, this institution will probably not be effective in short term period, and it is the preliminary condition for the success of the next steps, such as for example effective changes in the legal sphere or development of the road traffic supervision system. Moreover, required changes in the drivers' behaviour are the key determinant for improved road safety and all the stakeholders need to be involved in transition to safer Polish roads—road transport participants, but also police, local self-governments, etc. Without their acceptance and commitment, the system change can be slowed down or hampered. Unfortunately, in Poland public interest in the form of improved road safety not always outweighs private interest, which is visible also with reference to public services responsible for road safety. For these reasons, meeting the goal of halving the number of road fatalities in Poland and improving Polish road safety statistics is still an open question. But, on the other hand, explicit and clear actions have been finally undertaken in order to build a coherent road safety system.

References

1. Global Plan for the Decade of Action for Road Safety 2011-2020, UN Road Safety Collaboration (2011), http://www.who.int/roadsafety/decade_of_action/plan/en/.
2. García-Altés, A., Pérez, K,: The economic cost of road traffic crashes in an urban setting. Inj Prev. 13(1), 65–68 (2007), http://www.ncbi.nlm.nih.gov/pmc/articles/PMC2610566/.
3. Social and economic consequences of road traffic injury in Europe. European Transport Safety Council, Brussels (2007), http://etsc.eu/wp-content/uploads/Social-and-economic-consequences-of-road-traffic-injury-in-Europe.pdf.
4. Social Costs of Accidents in Sweden, Swedish Civil Contingencies Agency (MSB) (2006), https://www.msb.se/RibData/Filer/pdf/26421.pdf.

5. René, D.: International Encyclopedia of Comparative Law. Brill Archive (1986).
6. Wycena kosztów wypadków i kolizji drogowych na sieci dróg w Polsce na koniec roku 2013 (Estimation of costs of road accidents and collisions in Poland at the end of 2013). Instytut Badawczy Dróg i Mostów na zlecenie Krajowej Rady Bezpieczeństwa Drogowego, Ministerstwo Infrastruktury i Rozwoju, Warszawa (2014), http://www.krbrd.gov.pl/files/file/Koszty-wypadkow-i-kolizji-drogowych-w-Polsce-w-2013-r_.pdf.
7. Bliss, T., Breen, J.: Road Safety Management. Capacity Reviews and Safe System Projects Guidelines. Updated Edition. Global Road Safety Facility, Washington (2013), p. 34, http://siteresources.worldbank.org/INTTOPGLOROASAF/Resources/2582140-1371235287855/Road-Safety-Management-Guidelines-EN.pdf.
8. Commission Communication to the Council for an action programme on road safety [COM (93) 246 final].
9. Reducing the Severity of Injury Through Post Impact Care. ETSC, Brussels, Belgium (1999).
10. Communication from the Commission to the Council, the European Parliament, the Economic and Social Committee and the Committee of the Regions: Promoting road safety in the European Union: the programme for 1997-2001 [COM (97) 131 final].
11. European Transport Safety Council (ETSC), http://etsc.eu/about-us/.
12. White Paper. European transport policy for 2010: time to decide [COM(2001) 370 final]. Commission of the European Communities, Brussels (2001).
13. Communication from the Commission on the European Road Safety Action Programme. Halving the number of road accident victims in the European Union by 2010: a shared responsibility [COM(2003) 311 final].
14. Road Safety. Mobility and Transport, European Commission, http://ec.europa.eu/transport/road_safety/specialist/erso/index_en.htm.
15. PIN. The Road Safety Performance Index. ETSC http://etsc.eu/projects/pin/.
16. Commission Communication of 20 July 2010 to the European Parliament, the Council, the European Economic and Social Committee and the Committee of the Regions – "Towards a European road safety area: policy orientations on road safety 2011-2020" [COM(2010) 389 final].
17. White Paper. Roadmap to a Single European Transport Area – Towards a competitive and resource efficient transport system [COM(2011) 144 final]. Commission of the European Communities, Brussels (2011).
18. Road safety statistics at regional level. EUROSTAT (2014), http://ec.europa.eu/eurostat/statistics-explained/index.php/Road_safety_statistics_at_regional_level.
19. Mobility and transport Newsletter 170 (2014), http://ec.europa.eu/transport/newsletters/2014/04-04/newsletter-print_en.htm.
20. EU Road safety policy. Mobility and Transport, European Commission (2015), http://ec.europa.eu/transport/road_safety/specialist/policy/index_en.htm#1.
21. Mid Term Review of the European Commission's Road Safety Policy Orientations 2011-2020. Briefing – Stakeholder Consultation Workshop, 17 November 2014. European Transport Safety Council (2014), pp. 2-4, http://etsc.eu/wp-content/uploads/2014-11-rspo-midterm-review.pdf.
22. Projects financed by the Commission. Road Safety. Mobility and transport, European Commission, http://ec.europe.eu/transport/road_safety/specialist/projects/sorted-by-domains/index_en.htm.
23. EU transport in figures. Statistical pocketbook 2014. European Union, Belgium (2014), http://ec.europa.eu/transport/facts-fundings/statistics/ pocketbook-2014_en.htm.
24. Road Safety Vademecum. Road safety trends, statistics and challenges in the EU 2010-2013. European Commission, DG for Mobility and Transport (2014), http://ec.europa.eu/transport/road_safety/pdf/vademecum_2014.pdf.
25. How safe are your roads? Commission road safety statistics show small improvement for 2014. European Commission - Press release, Brussels, 24 March 2015, http://europa.eu/rapid/press-release_IP-15-4656_en.htm.

26. Goniewicz, M., Goniewicz, K.: Wypadki drogowe w Polsce – czynniki sprawcze i zapobieganie (Road accidents in Poland – causes and avoidance). Bezpieczeństwo Pracy - Nauka i Praktyka 9 (468), 14-17 (2010).
27. Symon, E.: Wypadki drogowe w Polsce w 2013 roku (Road accidents in Poland in 2013). Komenda Główna Policji, Biuro Ruchu Drogowego, Zespół Profilaktyki i Analiz, Warsaw (2014).
28. Symon, E.: Stan bezpieczeństwa ruchu drogowego w Polsce w 2014 roku (Road safety performance in Poland in 2014). Komenda Główna Policji, Biuro Ruchu Drogowego, Zespół Profilaktyki i Analiz, Warsaw (2015).
29. Informacja o wynikach kontroli bezpieczeństwa ruchu drogowego w Polsce (synteza ustaleń i wnioski) (Information on results of the control of road safety in Poland (synthesis of findings and conclusions)). Najwyższa Izba Kontroli, Departament Komunikacji i Systemów Transportowych, Warsaw (2011).
30. NIK o przyczynach wypadków drogowych (NIK about the reasons for road accidents). Najwyższa Izba Kontroli (2011), https://www.nik.gov.pl/aktualnosci/nik-o-przyczynach-wypadkow-drogowych.html.
31. Road building in Poland. The facts and the myths, experience and perspectives. PwC, http://pzpb.com.pl/newpzpb/wp-content/uploads/Road-building-in-Poland_ver_ang.pdf.
32. Informacja o wynikach kontroli. Bezpieczeństwo ruchu drogowego (Information on the results of control. Road safety). Najwyższa Izba Kontroli, Departament Komunikacji i Systemów Transportowych, Warsaw (2014).
33. UCHWAŁA Nr 25 RADY MINISTRÓW z dnia 1 lutego 1972 r. w sprawie inspekcji gospodarki samochodowej (Resolution no. 25 of the Ministry Council of 1.02.1972 on the inspection of car management). Monitor Polski Nr 8, poz. 50.
34. Gerondeau, C.: Road Safety in Poland. Study of road safety in Central and Eastern Europe Countries. Report of Expert's Group. World Bank European Community, Paris (1992).
35. National Road Safety Council (Krajowa Rada Bezpieczeństwa Ruchu Drogowego, KRBRD), http://www.krbrd.gov.pl/en/.
36. Krajowy Program Poprawy Bezpieczeństwa Ruchu Drogowego w Polsce GAMBIT 2000 (National Programme for Improving Road Safety in Poland GAMBIT 2000), Warsaw (2000), http://v000384.home.net.pl/gambit/gambit2000.pdf.
37. Krajowy Program Poprawy Bezpieczeństwa Ruchu Drogowego na lata 2005-2007-2013 – GAMBIT 2005 (National Programme for Improving Road Safety 2005-2007-2013 GAMBIT 2005), Warsaw (2005), http://v000384.home.net.pl/download/pdf/Gambit_2005_RM.pdf.
38. Polityka transportowa Państwa na lata 2006-2025 (National Transport Policy for 2006-2025), Ministerstwo Infrastruktury, Warsaw (2005), http://archiwum-ukie.polskawue.gov.pl/HLP%5Cfiles.nsf/0/6EDB2FBCDE37665EC1257266004241F6/$file/transport.pdf.
39. Stomma, P.: Odpowiedź podsekretarza stanu w Ministerstwie Transportu - z upoważnienia ministra - na interpelację nr 2287 w sprawie planów rządu mających na celu poprawę stanu bezpieczeństwa ruchu drogowego (The answer of the vicesecretary of the state in the Ministry of Transport – on the behalf of the minister – to the interpellation 2287 related to the government plans aimed at improving road safety). Warsaw (2006), http://orka2.sejm.gov.pl/IZ5.nsf/main/74EC5E95.
40. Narodowy Program Bezpieczeństwa Ruchu Drogowego 2013 – 2020 (National Road Safety Programme 2013-2020). Krajowa rada Bezpieczeństwa Ruchu Drogowego (2013), http://www.krbrd.gov.pl/files/file/Programy/KRBRD-Program-P1a-20140422-S2-K3d-EN.pdf.
41. Strategia rozwoju Kraju 2020. Aktywne społeczeństwo, konkurencyjna gospodarka, sprawne państwo (National Development Strategy 2020. Active society, competitive economy, efficient state). Ministerstwo Rozwoju Regionalnego, Warsaw (2012), https://www.mir.gov.pl/english/Regional_Development/Development_Policy/NDS_2020/Documents/NDS%202020.pdf.
42. Program Realizacyjny na rok 2013 (Action Programme 2013). Krajowa Rada bezpieczeństwa Drogowego, Warsaw (2013), http://www.krbrd.gov.pl/pl/program-realizacyjny-2013.html.

43. Program Realizacyjny na lata 2014-2015 (Action Programme 2014-2015). Krajowa Rada bezpieczeństwa Drogowego, Warsaw (2014), http://www.obserwatoriumbrd.pl/resource/ff3b8a69-1307-4a88-a89c-d95aed10588f:JCR.
44. Centrum Bezpieczeństwa Ruchu Drogowego, Instytut Transportu Samochodowego (Road Traffic Safety Centre at the Motor Transport Institute), http://www.its.waw.pl/link,CBR,2.html
45. Polskie Obserwatorium Bezpieczeństwa ruchu Drogowego (Polish Road Safety Observatory), http://www.obserwatoriumbrd.pl/en/.
46. Jamroz, K., Kastner, M.: Rozwój system bezpieczeństwa ruchu drogowego (The development of a road safety system). Transport miejski i regionalny 7–8, pp. 32-39 (2006).
47. Informacja o wynikach kontroli. Bezpieczeństwo ruchu drogowego (Information on the control results. Road safety). Najwyższa Izba Kontroli (2013).
48. Country Report on Poland. Road Safety Management Capacity Review. Report No. 78319-PL. World Bank, Washington (2013).
49. Jamroz, K., Michalski, L.: Analiza możliwości realizacji celów strategicznych bezpieczeństwa ruchu drogowego w Polsce (The analysis of the possibilities of realisation of the strategic road safety objectives in Poland). Prace Naukowe Politechniki Warszawskiej 96 (2013), preprint.
50. Road Safety Management Profile. Poland. DaCoTa, Project co-financed by the European Commission, Directorate-General for Mobility and Transport (2010), http://ec.europa.eu/transport/road_safety/specialist/knowledge/dacota/pdf/rsm/dacota-rsm-profile-template-po_en.pdf.
51. Poland. Partner countries. EURORAP, http://eurorap.org/partner-countries/poland/.
52. Preliminary Consultation Note on Establishment of the National Road Safety Lead Agency for Poland. World Bank (2013).

Public–Private Partnership and the Development of Transport Infrastructure in Poland: The Analysis of Critical Success Factors

Beata Zagozdzon

Abstract Efficient transport is an important determinant of the integration of the EU countries. TEN-T has been created by the European Commission to ensure the mobility of citizens, as well as fast and safe trading. However, the European transport infrastructure is characterized by large variations on the level of development that occurs in the East and West of the EU. In Poland, the infrastructure is technically unadjusted to the current transport needs and preferences. Modernization and development of infrastructure requires large capital expenditures. At the same time the public finance sector is overloaded, and the budget deficit is common. European countries and governments around the world are more likely to engage the private sector to finance infrastructure. The usual form of involvement of private investors is the public–private partnership (PPP). However, not all projects are successful. Achieving tangible benefits of PPP requires the creation of appropriate conditions conducive to its implementation.

The aim of this study is to identify and analyze critical success factors for PPP transport infrastructure projects in Poland. The research literature proves that there are many factors and they relate to various aspects and stages of the PPP implementation. The factors are classified in four groups: financial and economic, political and legal, technical, and social. The ones that have significant impact on transport infrastructure include: stable macroeconomic situation of the country, political support and commitment of the government, legal system, well-organized public consulting agencies, financial market availability, credible and experienced private investors. In Poland, the most important factors that are necessary for success in the implementation of PPP projects are: support of the government and governmental institutions, legal framework and regulations and the existence of adequate public advisory bodies. For the past 20 years, these conditions have not been fully met and formed a barrier to the development of the partnership. The consequence of the lack

B. Zagozdzon (✉)
Faculty of Transport and Electrical Engineering, Kazimierz Pulaski University of Technology and Humanities in Radom, 26-600 Radom, Poland
e-mail: b.zagozdzon@pr.radom.pl

M. Bąk (ed.), *Transport Development Challenges in the Twenty-First Century*, Springer Proceedings in Business and Economics,
DOI 10.1007/978-3-319-26848-4_12

of public advisory body is limited knowledge on PPP in public administration as well as in banks, which in the case of the latter, limits the availability of the financial market for private partners.

Keywords PPP • Critical success factors • Transport • PPP in Poland

Introduction

Creating a unified European space is one of the principal objectives of the European Union. Territorial cohesion concerns various areas, but efficient transport links play the key role. The multimodal TEN-T network aim to ensure full mobility of citizens as well as fast and safe trade. This network is supposed to improve the performance of supply chains, by the means of efficient transport corridors and logistics platforms.

At the same time, the European transport infrastructure varies in levels of development, especially in the east and west of the EU. The necessity to modernise and develop infrastructure is a serious problem in Poland, where the infrastructure is technically not adjusted to the current transportation needs and requires huge capital expenditures. The World Bank estimates that 65 billion EUR are needed for infrastructure investments in the new EU member states in the years 2007–2032. Poland is estimated to have the greatest needs—21.4 billion EUR [1]. At the same time, the public finance sector is overloaded and budget deficits are common. Therefore, the European countries and governments in the entire world are increasingly willing to engage the private sector to finance the construction and development of infrastructure. Public–private partnership (PPP) is the most common form of engagement among private investors.

Since the topic of PPP is quite well known and addressed to by numerous publications, the article does not cover a broader discussion on the nature and forms of partnership. It provides a general definition of PPP, formulated on the basis of its various definitions that can be found in studies on the subject [2, 3]. Public–private partnership is a type of agreement between the government and a private partner. The institutions cooperate when implementing the project. However, they have their own interests and try to achieve their separate goals. The partners cooperate using clearly defined allocations of risk and responsibilities. The cooperation should result in lower costs and higher quality of services than those originating in traditional financing from public funds [4]. The form of private partners' involvement in PPP projects depends on the specificity of the given project and the individual needs of project participants. The detailed classification of PPP forms have been presented by for example R. Mu, M. de Jong, and E. ten Heuvelhof [5]. At this point, the main forms of private partner's participation in infrastructure projects are worth mentioning. These include: BOT—Build, Operate, Transfer (to the public sector after a few years since the project was launched), DBFO—Design, Build, Finance, and Operate and BOO—Build, Own, Operate (in this case the private partner never stops to be the owner).

The most important area of the implementation of PPP projects is transport infrastructure. Since the 1990s, transport projects have had the highest share in the total value of the European PPP market. The funds allocated to these projects comprised on average 75–90 % of the total value of projects in 1995–2009 and reached 80 % in 2014 [6].

Public–private partnership has been the area of interest of the European Commission for about 20 years, especially in the context of the construction of TEN, which is evident in directives, communications, and studies—Guidelines for Successful Public–Private Partnerships [7], Resource Book on PPP Case Studies [8], A Guide to Guidance. Sourcebook for PPPs [9]. In the years 1996–2011, the EU allocated 3738 million euros to 11 PPP projects [10]. The combination of PPP with EU funds is also a prominent issue in the programming period 2014–2020. "White Paper. Roadmap to a Single European Transport Area—Towards a competitive and resource efficient transport system", a document published in 2011, states that "diversified sources of finance both from public and private sources are required", adding that "unlocking the potential of private finances equally requires an improved regulatory framework and innovative financial instruments" [11]. The Commission promotes using private capital to a much greater extent than it has been done so far.

The activities supporting PPP are important and needed, because the practice of implementing such projects shows that it is much more difficult to achieve benefits with no negative effects suggested by the subject literature [12]. Although PPP projects have been widely implemented in the EU since 1970, they are not the panacea for the increasing investment and financial needs, since the success of partnership requires creating appropriate economic, legal, institutional, and social conditions [13, 14].

The aim of the study was to identify and analyze the critical success factors for PPP for transport infrastructure projects in Poland. The issue is very important, because many problems have occurred and relatively few projects have been successfully implemented in the last 20 years of PPP implementation. In the years 2009–2013, only 21 % of the announced tenders ended with the signing of a contract. This leads to the following questions: Why are PPP projects more often implemented in other countries than in Poland? What factors determine the success of PPP projects? The discussion covers the concept of success very synthetically (because it is not the major area of the study) and refers to a PPP project that has been implemented in accordance with the budget, schedule, and the standard of service quality.

In attempt to answer the above questions, published materials on the subject was first consulted. They fostered an overall identification of PPP determinants. The success factors suggested by scholars were classified into four groups: financial and economic, political and legal, technical, and social. Then, factors that were most prominent for transport infrastructure projects were selected. The following stage of the study included the analysis of the critical success factors for PPP for transport infrastructure projects in Poland. The identification of these factors is important for the use of PPP, because it shows areas in which it is necessary to take action and optimally adapt it to the specificity of partnership. This creates conditions for a broad implementation of projects, which may accelerate the process of infrastructure modernization and full European integration.

Critical Success Factors for PPP: Review of Published Studies

Published studies on the subject suggest that conditions conducive to the implementation of PPP are defined as critical success factors (CSFs). Their identification was the subject of an extensive research conducted by Hardcastle et al. [15], who divided the factors into five main groups: effective procurement, project implementability, government guarantee, favourable economic conditions, and available financial market. The first group includes the following factors: transparent and competitive procurement process, government support, and the existence of good and competent public agencies, which help to implement PPP projects. The second group includes the following factors: a favourable legal framework, project technical feasibility, appropriate risk allocation, and a strong private consortium. Government guarantee means ensuring project revenues as typical business objectives are not always the main aim of these projects. In the remaining group of factors we can find, among others, the stable macroeconomic situation as well as the directions and priorities of the economic policy.

The classification of CSF, which is based on different aspects of risk associated with PPP projects, was developed by Ozdoganm and Birgonul [16]. Technical, financial and economic, social, environmental, and political and legal risks are important factors considered by decision-makers. The factors classified on this ground were divided into the following four groups:

- Financial and economic:
  - The need for services
  - Government guarantee
  - A strong team of members of the consortium
  - Project profitability
  - A stable economic environment in the country
  - The ability of the project to attract foreign capital
- Political and legal:
  - The stability of the political environment in the country
  - Government experience in implementing PPP projects
  - A transparent procurement system
  - A mature legal framework
  - Adequate regulations
- Technical:
  - The size of the project enabling participants to control it technically.
  - The reliability and experience of a private consortium.
  - Innovative solutions applied in the project.
- Social:
  - Public acceptance of the project.
  - The consistency of the project with requirements of environmental protection.
  - The price of future services.

A broad overview of CSFs was also provided by Ng et al. [17], who studied the importance of these factors in terms of expectations held by the public and private sectors as well as by the society. They grouped the factors into the following five groups:

- Technical:
  - The size of the project that can be technically manageable by a single consortium.
  - The possibility of applying innovative solutions.
  - Government experience with similar projects.
  - An experienced and reliable private consortium.
  - A flexible contract (enabling to change the specification).
- Financial and Economic:
  - The stable economic environment of the country.
  - Sound governmental economic policy.
  - A project that is more cost-effective compared to traditional forms of project delivery.
  - A project that can be self-funded to a significant extent.
  - A project that is financially interesting for the private sector.
  - A project that can attract foreign capital.
  - A project that is bankable, with limited completion from other projects.
- Social:
  - A long-term demand for the products/services in the community.
  - Understanding and support of the society.
  - Demand for services/products is stable and reliable.
  - An acceptable level of toll/tariff in the community.
  - A project that can create more job opportunities.
  - A project consistent with the natural environment protection policy.
- Political and Legal:
  - A favourable legal framework (mature, reasonable, and predictable).
  - Compliance with current regulations.
  - A stable political environment.
  - A project that is not politically sensitive, with political support for the project.
- Other Factors:
  - Compliance with government's strategic and long-term objectives.
  - Government support (e.g. guarantees or loans).
  - Fair conditions of the project to employees.
  - Support for the project and a commitment of employees of all parties to the project.
  - Flexibility in deciding on appropriate risk allocation.
  - Shared project management between public and private sectors,
  - Effective control mechanism over the private consortium.

The review enabled identification of critical success factors for PPP. They were classified into four groups: economic and financial, political and legal, technical, and social. Moreover, the factors were divided by the level of the analysis. The factors that are shaped by the government and its institutions are concentrated on the macro level; while those that influence practical implementation, at the development phase of the project and assessment of its feasibility, concern the micro level (Table 1).

Table 1 Critical success factors for PPP

	Economic and financial factors	Political and legal factors	Technical factors	Social factors
Macro-level factors	Favourable, stable macroeconomic conditions Government's economic policy supportive to PPP Availability of funding Transparent and competitive procurement process	Stable political environment Political support for PPP Favourable legal framework (mature, reasonable and predictable) Well-organized and competent public agencies Corruption control system and the rule of law	Government experience with similar PPP projects	Supportive and understanding society
Micro-level factors	Economic viability of the project Project is more cost-effective than traditional forms of project implementation Project is bankable and its profitability is sufficient to attract private investors and lenders Project can attract foreign capital Project can be self-funded to a significant extent or on a non-recourse basis	Project compatible with current regulations Project is not politically sensitive	Reliability and experience of a private consortium Size of the project enabling to manage it by a single consortium Possibility of innovative solutions Flexible contract (enabling to change the specification)	Long-term demand for the products/services Level of toll/tariff is acceptable Delivery of services/products is stable and reliable Project is environmentally sustainable Project can create more job opportunities

Source: Hardcastle, C., Edwards, P. J., Akintoye, A., Li, B.: Critical success factors for PPP/PFI projects in the UK construction industry: a factory analysis approach. In Proceedings: Public private partnerships e Opportunities and challenges, February 22, pp. 75–83. Hong Kong Convention and Exhibition Centre, Hong Kong (2005); Ozdoganm, I. D., Birgonul, M. T.: A decision support framework for project sponsors in the planning stage of build-operate-transfer (BOT) projects. Construction Management and Economics, No. 18(3), pp.343–353 (2000); Ng, S.T., Wong, Y.M.W., Wong, J.M.W.: Factors influencing the success of PPP at feasibility stage—A tripartite comparison study in Hong Kong. Habitat International, No. 36, pp.423–432. Elsevier (2012)

CSF Analysis of Major Importance for PPP

The factors identified above may vary in importance for the success of PPP schemes. Some of them have greater impact and their fulfilment is essential for a successful implementation. A study conducted by Chou et al. [18] shows that the following macro-level factors have the greatest influence: a stable macroeconomic environment, government's economic and political support, the availability of funding, the legal system, well-organized and competent public agencies, social support, a transparent procurement process, and a competitive procurement process. A study done by Ng et al. [17] indicates the most important micro-level factors. These are the following: a socially acceptable level of toll/tariff, a reliable and experienced private consortium, long-term demand for services, a stable and reliable delivery of services, project profitability that is sufficient to attract private investors, and a cost-effectiveness better than when compared to traditional forms of project delivery. In the next section of the article, only macro-level CSFs of the greatest importance for the project will be analyzed. The focus has been given to this group of factors, because they form the fundamental and basic framework for partnership. If they are not fulfilled, application of PPP will never be possible, not to mention successful implementation.

Thus, according to Hammami et al. [13] and Chou et al. [18], an important factor for transport infrastructure projects is the country's stable macroeconomic environment, in particular, indicators such as GDP growth, customers' purchasing power, and market size. Good macroeconomic conditions can attract investors and increase the level of funding [19]. This factor is of particular importance for developing countries that want to attract foreign private investors. Customers' purchasing power and market size, in turn, influence the level of tolls in the future, the volume of demand, and then the revenues and liquidity of a private consortium. As for large infrastructure projects, government guarantees (on financing and future demand) turn out to be an equally important factor.

As a result of a stable macroeconomic situation and government's political support, the partnership is more often applied, which, in turn, enables gaining experience in preparing and implementing projects. The studies made by P. Galilea and F. Medda [14] show that experience in PPP projects is one of the key CSFs. This refers to the experience of the government, private investors, and multilateral lenders in PPP projects. A reliable and experienced private consortium ensures the possibility to come to an understanding at the stage of preparing the project and its technical feasibility. The experience of financial institutions is equally important, because the private sector relies mainly on external financing when undertaking private-public partnership projects. Efficient financial intermediation, unrestricted access to the capital, and financial services are the most important factors for the success of PPP projects. Multilateral lenders' and credit agencies' involvement in the project reduces the risk of failure [20].

The consequence of the government's economic policy is a legal and institutional system. Mature and favourable legal regulations and the high quality of institutions are the primary CSF for PPP projects. According to the analysis carried out for the European Commission [8] and the studies done by Medda and Carbonaro, legislation is the first step to effective use of PPP [21]. What is important for investors is the stability of contracts including stipulations about the project. Therefore, the quality of institutions that make sure the law is obeyed is of huge prominence. According to studies, the partnership is more common in countries with strong and effective legal institutions as well as in countries where the rights of investors are well protected by regulations [13]. This creates a good climate for investors and has a positive influence on the trust between the government and private entities. Mutual trust is essential for the success of PPP schemes [22].

In the case of transport infrastructure, well-organized public PPP agencies are the key CSF. The studies made by Chou et al. indicate that it is the most important factor for transport infrastructure projects [18]. Advisory activity has particular significance for developing countries, which have no experience in applying PPP, and it is supposed to help these countries to overcome the trap of underdevelopment. It is also important that the agencies should be partially, if not entirely, financed from public funds [8]. Promoting PPP and spreading knowledge in the field of PPP helps in understanding this concept, especially its objectives and outcomes, and this, in turn, helps to gain support in the community.

The factor that is associated with government's economic policy and the legal system is a transparent procurement process and the level of competition in the public procurement market. The intensity of competition means a high number of offers being made, which offers the opportunity to select a better supplier and conduct ex-ante evaluation.

The Results of the Analysis and Implications for Poland

Polish experiences of cooperation between public and private sectors in financing and implementing transport infrastructure projects date back to 1994. In that year, the government passed the law that set the conditions for the preparations of the construction of motorways, the concession rules, and concluding contracts for the construction and operation of toll motorways. In the concession system, private consortia completed the following projects:

- The construction and use of Świecko-Konin section of A2 motorway of 255 km.
- The construction and use of Gdańsk-Toruń section of A1 motorway of 152 km.
- The construction by means of adjustment for the purpose of collection of tolls and use of Katowice-Kraków section of A4 motorway of 61 km.

Altogether, private consortia have constructed and operated approximately 20 % of the total length of the motorway network in Poland. The projects were initiated by the government in the years 1996–1997, and the roads were put into operation in the years 2004, 2008, and 2011/2012. In 2009, the government signed a concession contract with a private consortium to build and operate the motorway section of 180 km length. However, the consortium did not obtain funding from commercial banks and the concession expired. As a result, the motorway was built in the traditional system. The reason behind the failure of the project was the lack of consent from the government to increase the level of financial guarantees, and the global financial crisis, which resulted in a stiffened banks' position on granting credits for investments that carry a relatively high risk. As a result of problems that occurred when attempting to implement the project, the current transport policy of the government concerning financing transport infrastructure involves the use of mainly budgetary and EU funds. Projects that secure such a financing structure are easier to implement than those with the participation of a private sector.

The smaller PPP projects market, in turn, began to develop in 2009. These projects have been initiated by lower-level authorities—provincial, city, and municipal. They are implemented mainly in the following sectors: ICT infrastructure, sport and recreation, waste water and sewage, and municipal and transport infrastructure. In the years 2009–2013, the total number of 58 contracts with private partners were concluded, while 46 % of them were contracts worth up to 1.2 million euros, 19 % were contracts with the value of 1.2–4.8 million euros, 20 % were worth 4.8–12 million euros, and 14 % were contracts over 12 million euros. The total value of contracts signed during this period amounted to 720 million euros, while the average value of a contract concluded in 2013 was 8 million euros [23]. It is worth mentioning here that the value of PPP contracts (with financial closure) in the EU in the years 2009–2013 amounted to approximately 68 billion euros, while the average value of a single transaction was about 203 million euros in 2013 and 191 million euros in 2005–2014 [6]. These figures are given by European PPP Expertise Centre (EPEC), which considers in its studies the contracts at the level of no less than 10 million euros. The comparison of this data shows that the value of all PPP transactions in Poland, including contracts the value of which was less than 10 million euros, accounted for approximately 1 % of the EU market, while the value of contracts in Great Britain was about 50 %. Furthermore, the value of individual projects is low, and about 86 % of all contracts are worth less than 10 million euros. The PPP market in Poland is dominated by low-value projects, which are initiated by local, city authorities. Moreover, since 2009, the government has taken no action to initiate large transport infrastructure projects.

Previously conducted study on subject literature enabled the authors to identify a number of PPP critical success factors of varying impact on transport infrastructure projects. The most important ones in the group of macro-level factors are the following: a stable macroeconomic environment, government's political support, mature and favourable regulations, the availability of financing, and well-organized public agencies. The good economic condition of the economy, government's active support for partnership, and appropriate regulations form the foundation for the development of PPP. The next step is to initiate preparation processes and to implement PPP projects, which enables gaining experience. This is of great importance, because experience of the government, private investors, and multilateral lenders are CSFs that are significant for transport infrastructure projects.

In order to identify CSFs of PPP in Poland, the following procedures were used: the study done among public administration employees and private investors [24], analysis and reports prepared at the request of the Ministry of Economy [23, 25], reports prepared by consulting companies involved in PPP [26], and the report of the Supreme Audit Office (institution of public administration) [27].

Studies and expert analysis show that the group of the most important CSFs for implementation and development of PPP projects in Poland includes the following: the support of the government and its institutions, the legal system, the activity of public agencies providing advisory services, and the availability of funding [28, 29]. These factors coincide with the determinants that have been previously determined on the basis of study of published materials on the subject. The synthetic analysis of these factors will first cover the legal system and then the activity of advisory agencies and the availability of funding. These are macro-level factors, and they are dependent on the policy and actions that are undertaken by the government. Therefore, their analysis and the assessment of their fulfilment will enable determining the support of the government and its institutions for partnership.

Predictability of legal procedures and clear interpretation of law are important and crucial conditions for the implementation of PPP. In Poland, the first act that made it possible to implement road projects by the private sector was established in 1994. The Act, however, did not provide the possibility to provide financial support to private entities from the state budget, which was required by foreign institutions co-financing motorways. Only after 6 years, in 2000, the Act was amended and an additional source of financing introduced. In addition, the National Motorway Fund was established. It received budget grants for construction and funds from the state budget for guarantees and sureties for the concessionaire that was granted a bank credit. Another amendment to this Act was introduced in 2004, when the National Road Fund was established. It took over the responsibilities of the previous Fund. The first Act that offered the possibility to use PPP by lower-level authorities—provincial, county, and city—was introduced no earlier than in 2005. However, it formalized the way of carrying out the analysis of the project at the pre-implementation stage so much that no investment was implemented when it was in force. A new, more flexible act, which is currently in force, was adopted after three years, in 2008. The Act triggered real implementation of different forms of PPPs. The market of smaller PPP projects, excluding motorways, started to develop no earlier than in 2009.

Important problems, mentioned earlier, are differences in the interpretation of the valid law. There is no common understanding of certain concepts and expressions included in the regulations, even at the level of ministries of finance, infrastructure, or the economy. For years, local authorities have been waiting for a rigid interpretation of regulations governing the involvement of budget funds in PPP projects. It concerns mainly the assessment of the influence of liabilities arising from PPP contracts on the level of public entities' debt. The possibility to finance investments without burdening the limits of liabilities is one of the basic premises of the public entities' involvement in PPP contracts [23].

Poland is the one of few EU countries where there are no PPP public agencies. There is no institution that helps to assess the benefits of choice—whether a PPP or traditional formula should be used. It also concerns the development of standardized procedures, models of good practices and assistance in preparing and evaluating PPP project feasibility. Additionally, Poland lacks an extensive training program for the employees of public administration, private sector, and banks. As a result, the lack of knowledge in the field of PPP principles is considerable, and, in practice, it seriously hampers the development of partnership. The situation is due to a simple, existential mechanism: "What is unknown causes fear, anxiety, and reluctance." Limited knowledge of PPP leads to further difficulties in the implementation of projects. The public party is afraid that projects will be seen through the prism of "corruption links between business and politics", which means that it aims to avoid any political risk. The other party, a private investor, also fears accusations of corruption and the lack of transparency in the ways in which such projects are implemented. Furthermore, it is apprehensive of unwanted inspections and political turbulence at the local level, which can jeopardize the project.

Additionally, the low level of knowledge does not favour the involvement of banks in financing small projects prepared at local levels. Poland has a well-developed network of co-operative banks that according to experts are capable of financing small-scale projects. However, the lack of knowledge on the subject of full specificity of PPP contracts and banks' reluctance resulting from this lack turn into enormous obstacles. Therefore, in the opinion of private investors the problems with the bankability of projects form a significant barrier to their involvement in PPP projects.

As far as big infrastructure projects are concerned, availability of long-term financing is limited to a few institutions. The vast majority of banks do not presently accept the specificity of PPP projects. Even the most active ones, e.g. PKO Bank Polski, generally refuse to participate in the early stages of project preparation and wait with the credit decision until the final version of the contract is ready. The more common practice is the conservative attitude and refusing to grant a credit a priori in the case of projects where the risk of future revenues is on the side of a private investor. The lack of knowledge in the field of PPP specificity is not the only problem here. Another is the weak motivation for giving higher risk credits. Banks have huge demand for easy, short-term credits, with low risk so that they are not particularly interested in PPP contracts. The banks that are willing to finance PPP have other problems. These include the small number of large projects (a result of the current structure of the market, which is dominated by small projects). International financial institutions that have a fairly strong position on the market are interested in projects that have a value of over 24 million euros (more than 100 million PLN), because smaller projects are not profitable enough. To conclude, small banks do not have full knowledge on the specificity of PPP, which discourages them from becoming involved in such projects, while large financial institutions do not have projects that are worthy of their attention.

The practice of using PPP in Poland had exposed problems that were not regulated by law, and the period of change and development of appropriate legal solutions has lasted a few years. The underdeveloped legal system is undoubtedly a key obstacle to the partnership. Problems also occur when interpreting the existing regulations. There is no common understanding of concepts and expressions included in the regulations. Moreover, the lack of public advisory agencies hampers the development of partnership, which translates into limited knowledge among the low-level public administration, smaller banks, and private investors. Qualifications and skills in the field of PPP are already being obtained in Polish public, governmental, and self-governmental institutions. The problems are a consequence of insufficient government action. Experts underline and stress the lack of governmental policy aimed at using this formula of public services implementation (where, when and how), not applying PPP projects in the sectors within the competence of the government, even so obvious and necessary as roads and motorways. An important CSF support from the government and its institutions is not fully met.

Conclusions

The key CSFs for PPP transport infrastructure projects in Poland are consistent with the factors determined in the studies done by researchers in other countries. Experience in PPP gained in transport infrastructure projects in the world should be used in Polish projects. The most important for the successful implementation of PPP are macro-level factors: the support of the government and its institutions, the legal system, the existence and activity of public agencies providing advisory services, and the availability of funding. This is a group of basic, fundamental determinants, which are closely related to the priorities of the government's economic policy and government's actions. In Poland, these factors, despite their strong influence on the success of partnership, are not sufficiently met. The legal system is not fully adapted to the specificity of PPP projects, and appropriate legal solutions have been in preparation for a few years. In addition, the lack of a clear, generally applicable interpretation of regulations poses a huge problem. The government does not create public institutions dedicated to PPP, which would help to prepare and evaluate projects free of charge. Small government's engagement in promoting partnership results in gaps in knowledge among the employees of the public administration, private investors, and bank employees. The lack of knowledge is what experts refer to as the factor hampering the development of partnership, because it generates a conservative attitude and reluctance to enter into PPP arrangements. Projects are initiated by local authorities, while the government does not take any action even in typical PPP sectors, like motorways. So far, it has been easier to obtain EU funds than implement difficult, multilateral PPP projects. The situation does not help to gain experience, which is so important when implementing transport infrastructure projects.

The studies may be useful for decision-makers at various levels. They are responsible for creating conditions for the development of PPP. The knowledge in the field of CSFs that are the most relevant enables the government, local authorities, and PPP institutions to eliminate the key obstacles to partnership. In future studies, it would be worth conducting an in-depth analysis of CSFs in Poland, using statistical methods and econometric models. The analysis may be comparative in nature and may concentrate on Poland, the new EU member states (which joined the EU in 2004), and the existing research on an international scale. As a last resort, further research may focus on a constant update of CSFs for transport infrastructure projects.

References

1. Private Participation in the Transport Sector. Transport Papers, TP-24. World Bank, Washington (2009)
2. Flyvbjerg, B., Bruzelius, N., Rothengatter, W.: Megaprojects and Risk: An Anatomy of Ambition. Cambridge University Press, UK (2003)
3. Devapriya, K.A.K.: Governance issues in financing public–private partnership organisations in network infrastructure industries. International Journal of Project Management, No. 7, vol. 24, pp. 557–565 (2006)
4. Osborne, D., Gaebler, T.: Reinventing government – how the entrepeneurial spirit is transforming the public sector. Plume, New York (1993)
5. Mu, R., de Jong, M., ten Heuvelhof, E.: A Typology of Strategic Behaviour in PPPs for Expressways: Lessons from China and Implications for Europe. EJTIR, Issue 10(1) (2010)
6. Review of the European PPP Market. First half of 2014. EPEC (2014)
7. Guidelines for Successful Public – Private – Partnerships. European Commission, Directorate-General Regional Policy, Brussels (2003)
8. Resource Book on PPP Case Studies. European Commission, Directorate-General Regional Policy, Brussels (2004)
9. A Guide to Guidance. Sourcebook for PPPs. EPEC, EIB (2010)
10. EU Funds in PPPs. Project Stocktake and Case Studies. EPEC (2012)
11. White Paper. Roadmap to a Single European Transport Area – Towards a competitive and resource efficient transport system. COM(2011) 144 final, Brussels (2011)
12. Hart, O.: Incomplete contracts and public ownership: remarks, and an application to public-private partnerships. The Economic Journal, No. 113, p.p.69-76 (2003)
13. Hammami, M., Ruhashyankiko, J.F., E. Yehoue, E.: Determinants of public-private partnerships in infrastructure. IMF, Working Paper WP/06/99 (2006)
14. Galilea, P., Medda, F.: Does the political and economic context influence the success of a transport project? An analysis of transport public-private partnerships. Research in Transportation Economics, No. 30, pp.102-109 (2010)
15. Hardcastle, C., Edwards, P. J., Akintoye, A., Li, B.: Critical success factors for PPP / PFI projects in the UK construction industry: a factory analysis approach. In Proceedings: Public private partnerships e Opportunities and challenges, February 22, pp. 75-83. Hong Kong Convention and Exhibition Centre, Hong Kong (2005)
16. Ozdoganm, I. D., Birgonul, M. T.: A decision support framework for project sponsors in the planning stage of build-operate-transfer (BOT) projects. Construction Management and Economics, No. 18(3), pp.343-353 (2000)
17. Ng, S.T., Wong, Y.M.W., Wong, J.M.W.: Factors influencing the success of PPP at feasibility stage - A tripartite comparison study in Hong Kong. Habitat International, No. 36, pp.423-432. Elsevier (2012)
18. Chou, J.-S., Tserng, H.P., Lin, Ch., Yeh, Ch.-P.: Critical factors and risk allocation for PPP policy: Comparison between HSR and general infrastructure projects. Transport Policy, No. 22, pp.36–48. Elsevier (2012)
19. Mansoor, D., Klein, M.: Government Support to Private Infrastructure Projects in Emerging Markets. Policy Research Working Paper, No. 1688. World Bank, Washington (2007)
20. Butkiewicz, J., Yanikkaya, H.: The effects of IMF and World Bank lending on long-run economic growth: an empirical analysis. World Development, 33(33), 371-391 (2005)
21. Medda, F., Carbonaro, G.: Public-private partnerships in transportation: Some insights from the European experience. EIB, Working Paper (2007)
22. Parker, D., Hartley, K.: Transaction costs, relational contracting and public private partnerships: a case study of UK defence. Journal of Purchasing and Supply Management. No. 3, vol. 9, pp. 97-108 (2003)
23. Raport rynku PPP. Fundacja Instytutu PPP, Kancelaria Doradztwa Gospodarczego CieŚlak i Kordasiewicz, Warszawa, Poland (2013) (Published in Polish); PPP market report, PPP Institute Foundation. Office of Economic Consulting CieŚlak and Kordasiewicz, Warsaw, Poland (2013)

24. Herbst, I., Jadach-Sepioło, A., Marczewska, E.: Analiza potencjału podmiotów publicznych i przedsiębiorstw do realizacji projektów PPP w Polsce. PARP, Warszawa, Poland (2012), www.ppp.parp.gov.pl (pobrano: 19.01.2015) (Published in Polish); Herbst, I., Jadach-Sepioło, A., Marczewska, E.: The analysis of the potential of public entities and companies to implement PPP projects in Poland. PARP, Warsaw, Poland (2012), www.ppp.parp.gov.pl (downloaded on 19.01.2015)
25. Raport – porównanie systemów wspierania PPP w wybranych krajach. Ministerstwo Gospodarki, Wydział PPP, Warszawa, Poland (2013), www.mg.gov.pl (pobrano 10.01.2015) (Published in Polish); The report - a comparison of PPP support systems in selected countries. Ministry of Economy, PPP Department, Warsaw, Poland (2013), www.mg.gov.pl (downloaded on 10.01.2015)
26. Hausner, J. (red.): Raport o partnerstwie publiczno-prywatnym w Polsce. Centrum PPP, Warszawa, Poland (2013) (Published in Polish); Hausner, J. (ed.): The report on public-private partnership in Poland. PPP Centre, Warsaw, Poland (2013)
27. Realizacja przedsięwzięć w systemie PPP. Informacja o wynikach kontroli. Najwyższa Izba Kontroli, Nr ewid. 1/2013/P/12/051/KGP, Warszawa, Poland (2013), www.nik.gov.pl (pobrano 15.01.2015) (Published in Polish); The implementation of projects in the form of PPP. Information on the results of an audit. Supreme Audit Office, Reg. No. 1/2013/P/12/051/ KGP, Warsaw, Poland (2013)
28. Zagozdzon, B.: Determinants of Implementation of Public-Private Partnership in Poland: the Case of Transport Infrastructure. Advances in Economics and Business, 1 (2), pp.57 - 71. DOI: 10.13189/aeb.2013.010203. Horizon Research Publishing (HRPUB), USA (2013)
29. Gołębiowski, G., Marchewka-Bartkowiak, K. (red.): Partnerstwo publiczno-prywatne. Studia BAS, Nr 3(39), Warszawa, Poland (2013) (Published in Polish); Gołębiowski, G., Marchewka-Bartkowiak, K. (ed.): Public-Private Partnership. BAS Studies, No. 3 (39), Warsaw, Poland (2013)

Transportation Services as Specific Logistics Projects for Oversized Cargo in Poland

Iwona Pisz and Iwona Łapuńka

Abstract Market observation demonstrates that the transportation-freight forwarding-logistics sector carries out diverse projects based on clients' orders. The majority of those orders are unique and project related. Recently, the project approach to logistics has gained significance. Logistics project management is a relatively new area of knowledge about logistics and supply chains. The paper presents the essence and characteristics of transportation services as a specific logistics project. The authors characterize the transportation-freight forwarding-logistics sector, with particular focus on the sector of oversized cargo transportation. The demand for such services depends on the industry, energy, infrastructure development, investment projects in particular countries, as well as on economic policies. The paper identifies key factors that need to be taken under consideration when planning oversized cargo transportation services, i.e., specific characteristics, conditions, technology, tools, techniques, and methods, for example, a method of transportation service cost estimations. The proposed method uses fuzzy set theory. It can be implemented into the practice of transportation-freight forwarding-logistics enterprises, facilitating the modeling of uncertainties typical in such undertakings. A practical example of a transportation service is presented.

Keywords Transportation-freight forwarding-logistics sector • Transportation service • Logistics project • Oversized cargo • Oversized load • Logistics service provider

I. Pisz (✉)
Department of Logistics and Marketing, Faculty of Economics, Opole University, Opole, Poland
e-mail: ipisz@uni.opole.pl

I. Łapuńka
Department of Project Management, Faculty of Production Engineering and Logistics, Opole University of Technology, Opole, Poland
e-mail: i.lapunka@po.opole.pl

M. Bąk (ed.), *Transport Development Challenges in the Twenty-First Century*, Springer Proceedings in Business and Economics,
DOI 10.1007/978-3-319-26848-4_13

Introduction

The basic service on the logistics market, arising out of the need for moving various goods from their collection point to their destination, is transportation. This includes transportation of special loads (abnormal loads, oversized cargo), which can be considered highly problematic and therefore requires special attention. The process of its management is similar to project management, especially logistics project management [1, 2]. A project involving logistics project is a multistage process and multifunctional effort, the activities of which depend on several people (enterprises), resources, and dependencies [3].

The services are the application of competences including knowledge and skills by one entity for the benefit of another [4, 5]. Logistics services are a unique subset of industrial services. This kind of services spans the boundaries between suppliers and customers and has become increasingly important to successful supply of chain operations. Logisticians understand that these sets of activities constitute the very essence of their business [6]. The logistics services can vary in complexity and can be carried out in short or long supply chains. Market observations indicate that logistics services have become increasingly sophisticated, by far exceeding their traditional perception [7]. They begin to resemble projects characterized by singularity, uniqueness, temporariness, limited budget, and, occasionally, innovativeness. The majority of commissions received by logistics service providers constitute separate and singular transportation-freight forwarding-logistics processes, which necessitate detailed analysis, planning, as well as appropriate management methods. Therefore, they are often treated as a specific type of projects, called logistics projects [8].

The aim of the paper is to present the essence and characteristics of transportation services as a specific logistics project. Authors characterize the transportation-freight forwarding-logistics sector, with particular focus on the sector of oversized cargo transportation. The demand for such services depends on the industry, energy, infrastructure development, investment projects in particular countries, as well as on economic policies. Transportation of nonstandard cargos creates nonstandard problems. The management of such services requires specific knowledge, resources, and managers. The paper identifies the key factors that need to be taken under consideration when planning oversized cargo transportation services, i.e., specific characteristics, conditions, technology, tools, and techniques.

Changes in the Transportation-Freight Forwarding-Logistics Business

Transportation-freight forwarding-logistics business enterprises should be analyzed in the context of processes carried out in the areas of supply, production, returns, and waste management. The functioning of enterprises in this line of business is determined by independent demand on the market of goods and services. Market research shows that the transportation-freight forwarding-logistics business is characterized by an increased intensity of competition, which makes traditional methods of managing such enterprises and of competing on the market suboptimal. Further observations indicate that the ability of adopting new solutions does not necessarily guarantee a long-lasting competitive edge of a given transportation-freight forwarding-logistics enterprise on the market. Under such conditions, swift and flexible adaptation to the needs of the client gains importance. This can be achieved by creating and implementing new and innovative ventures in order to improve the functioning of logistics, mostly by means of cooperation with other units, e.g., by carrying out logistics projects. Such undertakings allow optimization and cost reduction in a logistics chain, and thus entrepreneurs should perceive them as a main factor that facilitates achieving a competitive position [9].

With regard to supply, the observation of the market of logistics services provided by diverse transportation-freight forwarding-logistics business enterprises reveals the following ongoing transformations: an increase in the number of enterprises providing specialized logistics services; migrations between enterprises; a shift in the roles of transportation-freight forwarding-logistics business enterprises related to the introduction of new customer service strategies, new technologies, products, and services; change in the market share; as well as increasing competition. In economic practice, this means that changes in the functioning and development of transportation-freight forwarding-logistics business enterprises are perceptible on the market of goods and services. Large transportation-freight forwarding-logistics enterprises offer a broad scope of services on a large scale, offering their clients values bundled to maximize benefits. Moreover, they effect improvement by implementing measures that allow their clients to enhance their operational effectiveness, competitiveness, and innovativeness, as well as change the functioning of their supply chains.

The implementation of diversification strategies by large transportation-freight forwarding-logistics business enterprises facilitates the identification of market opportunities, the flexible reaction to clients' needs, and the reduction of risks associated with potentially worse financial condition of a given group of clients. Large transportation-freight forwarding-logistics business enterprises outsource transportation functions to small business units, focusing on the management of a large number of geographically dispersed subcontractors. In order to effectively carry out such processes, IT solutions are constantly developed and subsequently implemented in enterprises. The implementation of an IT system for the purposes of synchronizing the flow of goods constitutes an example of a logistics project execution. Practically speaking, in this case, the future success in logistics depends on possessing IT tools for the effective management of extensive logistics projects. Specialization is yet another trend discernible on the market of logistics services—here many enterprises sense their opportunity, focusing on carrying out highly specialized logistics services, where the competition from large businesses is comparatively small. This is, for instance, the case with specialized transportation, including oversized cargo, as well as with value-added services.

On the other hand, small- and medium-sized transportation-freight forwarding-logistics business enterprises strive to increase their transportation potential by purchasing specialized means of transportation and loading and unloading equipment. Such enterprises introduce new products and services into their offer and aim to perform new functions, e.g., freight forwarding, warehousing, and relocating equipment and machinery, in order to gain new clients and increase the satisfaction of their regular clients. They employ the strategy of focusing on a market niche and decide to cooperate with other enterprises through outsourcing more often than large enterprises. Increased diversification increases their chances of survival and development on the market and creates a basis for developing and increasing competitiveness and innovativeness. This is particularly noticeable in the way third-party logistics (3 PL) operators perform their functions, aspiring to the role of fourth-party logistics (4 PL) operators [7, 10]. Proper execution of tasks by such logistics operators requires particular competences, access to appropriate technologies, IT support, and know-how. At present, the management of workflow in transportation-freight forwarding-logistics business enterprises is a complex, unique, and innovative process. Therefore, it requires extensive knowledge, skills, and appropriate competences as well as proper management methods and procedures, including simulation and modeling of material flow management, and multivariate analysis of the effectiveness of logistics processes. Bearing this in mind, it is necessary to implement the project approach into management practice into transportation-freight forwarding-logistics business enterprises.

Contemporary transportation-freight forwarding-logistics business enterprises discern the necessity of building interorganizational relations with specialized logistics enterprises, including those small and medium sized, based on active or passive cooperation, and hence the need to create specialized clusters grouping together given enterprises in a particular line of business. Building relations between enterprises allows to integrate and connect processes and resources for the purposes of carrying out services, as well as to introduce added value for clients. The nature of interorganizational relations results from the strategy adopted by the enterprises. In order to carry out complex undertakings, including logistics projects, enterprises build partnerships with other participants in the transportation-freight forwarding-logistics market. Such integration constitutes a set of undertakings and initiatives, which are used by a given enterprise in order to integrate the processes with its suppliers and recipients. Such activities allow enterprises to carry out diverse investments and thus to expand the scope of services they currently provide; in so doing enterprises better adapt to the increasing expectations of their clients.

The increasing number of partnerships based on logistics outsourcing arises from the necessity for cooperation as part of carrying out a particular order, which exceeds the capabilities of a single transportation-freight forwarding-logistics business enterprise or results from undertaken investment projects. Cooperation as part of challenges undertaken by enterprises contributes to the development of more flexible organizations, which are based on knowledge, core competences, and mutually profitable long-standing business relations. This also promotes learning in organizations. In this context, a learning organization would be a transportation-freight forwarding-logistics business enterprise continually expanding its capabilities in order to shape its future. A learning organization is a place where people can continually learn, develop, and improve their qualifications. Such an organization is capable of reacting to market needs in a swift and effective manner, adapting to market requirements, and responding to market changes. Over the course of time, it can become an intelligent enterprise, which, by means of learning, reaches its optimal state, looks for development opportunities, and achieves successes, simultaneously avoiding setbacks.

An increasing number of enterprises implement the project approach into their business practice. Market observation demonstrates that transportation-freight forwarding-logistics business enterprises carry out diverse projects based on clients' orders. The majority of those orders are unique and project related. Recently, the project approach to logistics has gained significance. Logistics project management is a relatively new area of knowledge about logistics and supply chains. The increased interest in planning and carrying out logistics projects and the associated problems is mirrored by an increasing number of trainings on the market of advisory and training services, as well as by an increase in scholarly publications. The significance of project management is supported by the examples of projects that ended in success or failure, carried out in different lines of business and in a variety of enterprises, including transportation-freight forwarding-logistics business enterprises.

The market of logistics services is continually on the rise. The number of enterprises competing for the largest project-based orders increases as well. The analysis of financial data of logistics enterprises demonstrates an increase in the number of enterprises whose turnover exceeds 100 million zlotys. The number of such enterprises has doubled since Poland's accession to the European Union. At the same time, the group of enterprises capable of developing their sales at the pace of the leaders in the area, and achieving similar incomes, expands in a consistent manner. This segment stands out as a market entirety among a mass of smaller enterprises (acting as subcontractors), whose development is significantly slower and income lower by far. Passive and active cooperation established between transportation-freight forwarding-logistics business enterprises brings about not only notable benefits but also dangers. Without doubt, one of key benefits is the diversification of activity as part of currently developed networks of designated logistics operators and integrators.

Defining Oversized Cargo Transportation

Present studies do not provide a precise or uniquely applicable definition of the oversized cargo. The determinates of oversized cargo are [11]:

- Cargo dimensions
- Cargo weight
- Available cargo space on the vehicle
- Permissible pressure and stress on the loading surface
- Permissible stress on surface of road/rails

Table 1 Definition of oversized cargo in different modes of transport

Transport mode	Definition of oversized cargo
Road transportation	Oversized cargo exceeds maximal permitted parameters of standard road vehicle or exceeds permissible axle load of the vehicle. In consequence, there are oversized vehicles instead of oversized cargos
Rail transportation	Oversized cargo exceeds standard loading gauge or exceeds permissible axle load of the railway. Such a situation is called extraordinary delivery, which means that such transport can cause difficulties in rail transport and it is necessary to take special technical and/or operating actions
Inland shipping	Oversized cargo is a cargo that overcomes the vessel's length and/or width or which overcomes the standard air draft of the vessel (vertical clearness of bridges, gates, etc.). It is taken under consideration on the restricted visibility of the helmsman as well
Sea transportation	Oversized cargo is defined as: break bulk or general cargo unit which overcomes the parameters of standard cargo units. This means it weighs hundreds or even thousands of tons and its dimensions are counted in tens or even hundreds of meters
Air transportation	Oversized cargo is such cargo which cannot be located in air container or on special consolidation unit. The only way to transport it is to use a special transport airplane
Intermodal transportation	Oversized cargo is the cargo that exceeds the average permissible parameters of means of transport in terms of size, shape, and/or permissible pressure and stress on the loading surface of at least one means of transport

Source: adapted from [11]

Oversized cargo can be classified with regard to its outer dimensions, weight, and shape. The particulars of a given cargo determine the methods and means of its transportation, the equipment, the mode of mapping out the route and acquiring permits, and, in turn, the feasibility of carrying out a particular order. Oversized cargo can be classified into a number of groups: standard over-dimensional, special over-dimensional (bulky but not heavy), heavy but not bulky, simultaneously heavy and bulky, and long [11]. Table 1 presents the definition of oversized cargo in different transport modes.

Carrying out diverse investments or unique projects in a number of countries requires delivering appropriate elements, i.e., the objects of delivery. The elements necessary for carrying out projects often constitute nonstandard cargo, whose weight and/or dimensions exceed the carrying capacity and/or load space parameters of standard means of transportation and/or road infrastructure. On the market of goods and services, there exists a demand for the transportation of diverse cargos from different parts of the world, over a variety of distances (Fig. 1). The cargos differ, for instance, in shape, dimensions, or tonnage. A demand for the transportation of nonstandard cargo, which cannot be transported by standard means, is observed. Economic growth and associated investments in the public and private sectors contribute to an increase in nonstandard cargo transportation. The oversized freight market is fueled by significant energy project [12]. Oversized cargo transportation

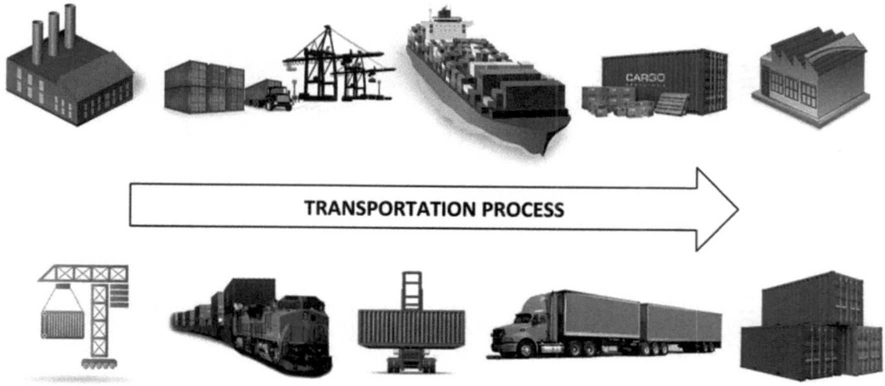

Fig. 1 An example of intermodal transportation of oversized load (source: own elaboration)

influences the economic development of a given country as well as the development of the industry, the infrastructure, and the energy sector.

A significant interest in nonstandard cargo transportation is observed on the market of goods and services. A smooth movement of the oversized cargo carrying vehicle needs special permits [11]. They are different in each country. In Poland, we can observe that there is a big demand for nonstandard transportation. Respectively, Table 2 and Fig. 2 present data on the number of permits issued by appropriate departments of the General Directorate for National Roads and Highways in Poland between 19 October 2012 and 31 December 2014. The data were obtained from individual departments of the General Directorate for National Roads and Highways at the request of the authors. According to the data, an increase in the demand for oversized cargo transportation is noticeable; this is particularly true with regard to very specific cargos requiring the issue of a one-time permit of category VII for the passage on national roads. It is estimated that oversized cargo transportation constitutes about 10 % of all transit [13].

The data presented in the table demonstrate that the number of permits issued for particular categories differs significantly between the departments of the General Directorate for National Roads and Highways. For instance, the departments in Wroclaw, Szczecin, and Poznan issue the highest number of category VII permits for the passage of nonstandard vehicles. In 2014, these departments issued over 45 % of all category VII permits. This may be due to the fact that the majority of oversized cargo transportation is carried out in those provinces (main communication routes) or that a large number of oversized cargo transportation enterprises are located in those provinces or in their vicinity. Moreover, the number of permits issued by a given department of the General Directorate for National Roads and Highways is influenced by the quality of service offered by these departments. It must be emphasized that entrepreneurs may apply for a permit of a given category in one of 16 departments. Surveys of oversized cargo transportation enterprises have shown that they usually apply for the category

Table 2 The number of permits issued for a given category for the passage on national roads in Poland issued appropriate departments of the General Directorate for National Roads and Highways in Poland for the period from 19 October 2012 to 31 December 2014

Headquarter of the department	Year 2012 Permit category	Year 2012 Permit category	Year 2012 Permit category	Year 2012 Permit category	Year 2013 Permit category	Year 2013 Permit category	Year 2013 Permit category	Year 2013 Permit category	Year 2014 Permit category	Year 2014 Permit category	Year 2014 Permit category	Year 2014 Permit category
	IV	V	VI	VII	IV	V	VI	VII	IV	V	VI	VII
Białystok	0	3	1	2	0	15	2	31	1	15	2	31
Bydgoszcz	3	48	6	15	3	138	42	232	14	105	39	232
Gdańsk	2	41	11	10	2	108	28	276	1	121	71	305
Katowice	17	52	30	25	15	99	86	235	25	100	107	181
Kielce	0	9	11	36	9	18	29	223	8	36	19	207
Kraków	0	22	10	45	20	101	60	112	11	94	49	217
Lublin	10	28	18	12	16	47	9	65	12	57	17	51
Łódź	4	30	14	7	16	48	21	35	14	58	36	48
Olsztyn	0	0	20	101	1	56	21	309	0	48	23	214
Opole	2	60	32	24	9	80	28	146	8	14	56	246
Poznań	7	103	37	184	28	288	99	770	29	309	167	494
Szczecin	2	45	13	40	3	127	22	279	3	127	31	502
Rzeszów	2	8	18	13	13	36	20	147	8	56	38	144
Warszawa	5	56	21	29	15	134	46	81	12	158	72	109
Wrocław	0	61	2	197	7	115	21	692	1	97	27	740
Zielona Góra	5	6	3	4	19	8	3	100	21	13	1	117
Total	59	572	247	744	176	1418	537	3733	168	1408	755	3838

Source: own elaboration

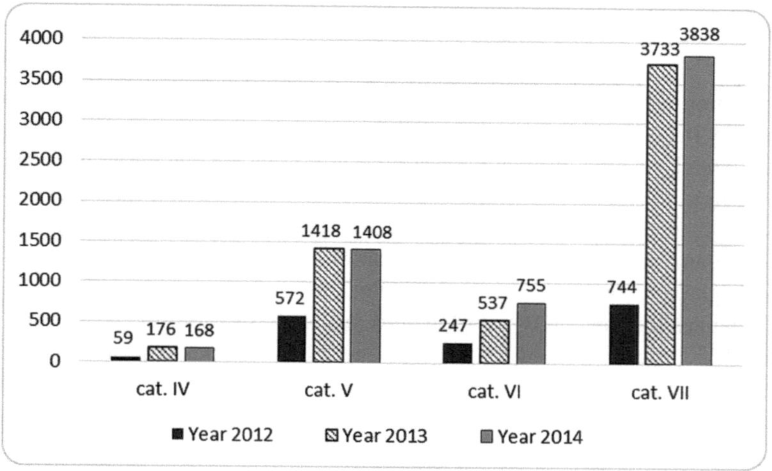

Fig. 2 Statistic of permits issued for IV–VII category for the passage on national roads in Poland for the period from 19 October 2012 to 31 December 2014 (source: own elaboration)

VII permits in the departments where service is most effective, the time needed to issue a permit is relatively the shortest, and the number of successful applications is high (the departments in Wroclaw, Szczecin, Poznan, and Warsaw). The situation is different for permits of other categories. In the case of category IV–VI permits, entrepreneurs usually apply in the departments of the General Directorate for National Roads and Highways in the province where a given enterprise is registered.

These are the actual data on the number of appropriate permits for nonstandard road transportation in Poland. Furthermore, a fraction of such transports are carried out without requisite permits. Sometimes, cargo parameters diverge from the ones specified in the permit, which means that entrepreneurs misdeclare the size of the transported cargo. The passage of nonstandard vehicles on public roads without permits or violating the conditions stipulated in the permit is subject to a fine under an administrative decision. Upon inspection, such a decision is issued by appropriate local unit of the police, road transportation inspection, border guards, customs, or the road administrator.

The data in Fig. 2 demonstrate that the number of issued category VII permits is on the increase. In this case, these are one-time permits for nonstandard vehicle passage on the route stipulated in the permit. They are valid for 14 days in the case of a single passage or for 30 days for multiple passages. Category IV–VI permits are issued for multiple passages and are valid for 1, 6, 12, and 24 months, respectively. The entrepreneur declares the required validity period when applying to the appropriate department of the General Directorate for National Roads and Highways for the permit for the passage of a nonstandard vehicle. Therefore, the analysis of the number of category IV–VI permits requires additional information on the validity period of the permits, which we were unable to obtain from the GDDKiA departments.

Oversized Cargo Transportation Planning

The planning process for moving oversized cargo continues to grow in detail and complexity (Fig. 3). Project cargo logistics is one of important aspects of the supply chain. As we have already stated, the service providers are becoming more sophisticated in how they plan and partner for success when transporting oversized and heavyweight cargo. Moving the oversized cargos is complex, especially with the permitting and planning necessary to transport this type of load [14]. A lot of planning is required to avoid problems of nonstandard load transportation. The enterprise which realizes this special type of logistics project has to control every detail of the project. This process includes many steps and should include what-if scenarios.

The basic tasks comprising oversized cargo transportation include: planning, budgeting, organizing (including obtaining appropriate permits), and supporting activities, i.e., pilotage, police escort, road infrastructure disassembly, raising railway/tramway systems, inspection of bridges and ferry loading ramps, tours of inspection, labeling vehicles according to local regulations, loading separate cargos onto ferries, and transshipment of oversized cargo. Of particular significance is the preparatory stage of a logistics project. Special requirements and conditions of oversized cargo transportation include various permits, route surveys from state departments of transportation, police escorts, etc.

The organization of oversized cargo transportation is a complex process, primarily consisting of: preparation of cargo for transportation, selection of appropriate means of transportation, route planning, and obtaining permits for oversized cargo transportation [15]. By analogy with logistics project planning, it is necessary to assess time, cost, and risk and to perform overall ex ante and ex post project evaluation.

Legal Basis for Oversized Cargo Transportation

Enterprises transporting elements with nonstandard features, shapes, or dimensions are required to carry out specialized services using nonstandard means of transportation and according to different procedures and regulations. Oversized cargo

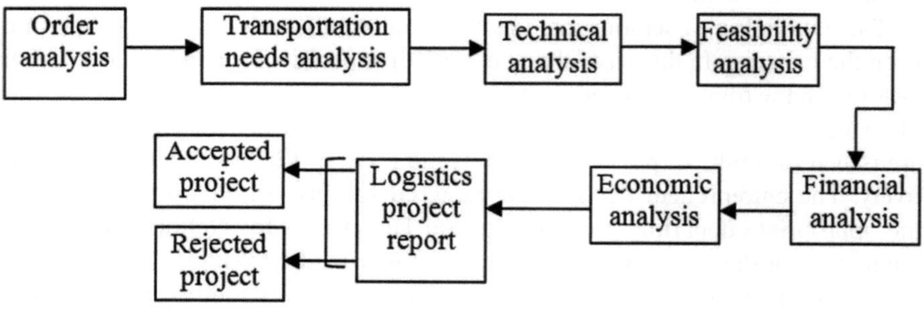

Fig. 3 An integrated model for decision analysis of logistics project execution (source: own elaboration)

transportation requires special permits, depending on the type of transportation and detailed arrangements with appropriate infrastructure administrators.

Oversized cargo transportation requires specific formal conditions to be fulfilled. Every country in which such transportation is carried out has its separate regulations with regard to permissible parameters. In practice, this involves permitted driving time, traffic regulations, conditions of pilotage, and police escort. Every country is different with regard to the conditions of securing special permits; and the waiting periods vary as well, ranging from several days up to several weeks. The appropriate authority or institution may refuse to issue a transit permit on a number of grounds, e.g., environment protection or lack of proper infrastructure. Permit prices vary depending on the country in which transportation is carried out and are related to the length and weight of the set, route length, number of axes, and length, width, and height of cargo and transport. Every special permit specifies the conditions of carrying out the transportation, which mostly concern speed limits and mandatory stopovers at certain locations. The transportation period (day or night) is also specified, along with an hourly time frame and the necessity for escort (if applicable). At times, bridge strength and capacity may need to be assessed or the route may require a tour of inspection.

The oversized cargo is transported mostly by road, because it is treated as the cheapest way and the most flexible means of transportation [11]. Transit of oversized cargo on public road requires special permits. In the case of international transit, such permits must be issued by each country en route. Moreover, every country has designated public administration units, which issue permits for oversized cargo transportation. In Poland, new regulations with regard to oversized cargo transportation have been in force since 19 October 2012. According to these regulations, nonstandard vehicle traffic is allowed as long as the following four conditions are fulfilled: a permit for the transit of a nonstandard vehicle of appropriate category has been obtained (such a permit must be issued based on an administrative decision of the appropriate authority); conditions of transit specified in the permit must be observed; the nonstandard vehicle must be piloted if one of the following is exceeded (length, 23 m; width, 3.2 m; height, 4.5 m; and gross vehicle mass, 60 ton); and the operator of a nonstandard vehicle must act with due caution. In addition, section 64a–64i of the act divides nonstandard vehicle transit permits into seven categories, depending on the technical features of the vehicle and the type of road on which transit takes place [16, 17]. Table 3 presents the basic characteristics of such permits.

The abnormal transportation business is a specific one. The companies which do the business deserve special attention. Usually they are gathered in some association. For example, ESTA is a European association of professional organizations representing companies active in the mobile crane rental and nonstandard road transportation business. ESTA represents national trade associations whose members are engaged in mobile crane and access platform rental as well as nonstandard road transportation companies. Individual companies can become special members, if there is no appropriate national association in their country. The primary objective of ESTA is the harmonization of rules and regulations for crane companies and haulers across Europe [18].

Table 3 The basic characteristics of permits of road transportation in Poland

Permit	Oversized vehicles	Type of road	Term of permit
Cat. IV	(a) GVW not higher than permissible values (b) Width not exceeding 3.4 m (c) Length not exceeding: • 15 m for a single vehicle • 23 m for a combination of vehicles • 30 m for a combination of vehicles with steering axles (d) Height not exceeding 4.3 m (e) Load axles not exceeding values designated for roads with permissible load of a single drive axle of up to 11.5 t	National	1, 6, 12, 24 months
Cat. V	(a) Axle loads not higher than values permissible for the specific road (b) Width not exceeding 3.4 m (c) Length not exceeding: • 15 m for a single vehicle • 23 m for a combination of vehicles • 30 m for a combination of vehicles with steering axles (d) Height not exceeding 4.3 m (e) GVW not exceeding 60 t	Public	1, 6, 12, 24 months
Cat. VI	(a) Width not exceeding: • 3.4 m for an undivided road • 4 m for a dual carriageway, LOS: A (motorway), S (expressway), and GP (fast traffic trunk road) (b) Length not exceeding: • 15 m for a single vehicle • 23 m for a combination of vehicles • 30 m for a combination of vehicles with steering axles (c) Height not exceeding 4.3 m (d) GVW not exceeding 60 t (e) Axle loads not higher than values designated for roads where the permissible single axle load does not exceed 11.5 t	National—in accordance with the list of roads enumerated in Art. 64c section 8	1, 6, 12, 24 months
Cat. VII	(a) Dimensions and GVW exceeding the values enumerated in categories I–VI (b) Axle loads higher than values designated for roads where the permissible single axle load does not exceed 11.5 t	The specific route indicated in the permit	14 days for single passage, 30 days for multiple passages

Source: adapted from [16]

One of the members of ESTA is the Polish Heavy Transport Association (OSPTN). The primary objective of OSPTN is to facilitate changes in legislation regarding nonstandard vehicle cargo transportation and nonstandard vehicle pilotage. OSPTN strives to introduce changes in the nonstandard vehicle registration and in the rules that govern the issuing of transit permits. One of its other aims is to facilitate changes in the so-called EU nonstandard vehicle certification of approval. Moreover, the association carries out scientific and technical activity with regard to nonstandard vehicle transit and supports social initiatives for constructing roads and expressways. In order to effectively carry out its goals, OSPTN cooperates with local governments and administrations, academic and cultural institutions, and non-governmental organizations. It also cooperates with business entities, public figures, and mass media. OSPTN carries out informational and promotional activity via a self-published magazine, a website, and an information bureau. It works for the general public by organizing conventions, workshops, meetings, and promotional actions; coordinating trainings and conferences devoted to transportation; as well as exchanging experiences with similar institutions abroad [19].

Organizing Logistics Project Resources and Activities

Carrying out a logistics service in the context of oversized cargo transportation displays features of a logistics project. It constitutes a group of diverse activities, the aim of which is to meet requirements in terms of time, cost, and scope. Oversized cargo transportation is the final element in a long chain of specialist logistics operations. It requires an appropriate approach, the use of suitable methods and equipment. From the perspective of an employer, i.e., an entrepreneur responsible for delivering oversized cargo, an order to deliver such cargo is a unique undertaking, which requires an individual approach. This necessitates an assessment of the market of oversized cargo transportation, selecting a logistics operator based on the chosen assessment criteria, and commissioning the organization and execution of oversized cargo transportation. In order to carry out such a logistics service, a contractor, i.e., a logistics operator, must prepare thoroughly, possess or rent appropriate equipment, as well as demonstrate knowledge and experience in this type of transportation. It is essential to determine the feasibility of transportation, analyze all possible means of transportation, choose the optimal one, and to assess and select loading and unloading equipment. Moreover, it is necessary to analyze available means of transportation and all permissible routes of oversized cargo transportation and to select the optimal route. A logistics operator should make plans for loading, unloading, and securing individual elements of the cargo and determine the possible approaches for protecting the cargo against external conditions.

Carrying out oversized cargo transportation requires estimating the time and cost of execution of this logistics service, performing risk assessment of the entire undertaking, an effectiveness assessment for a given project, and, finally, assessing and choosing the cooperators required for the efficient execution of a given order. To this end, it is necessary to prepare a work breakdown structure. Such structure enumerates the basic tasks, based on which the cost, time, and risk of a particular logistics project are assessed; moreover, it facilitates identifying the need for outsourcing services and selecting potential subcontractors for a project-based order. It is necessary to plan the time frames and effectiveness of individual activities and time correlations and to synchronize with other activities within the given enterprise and with its cooperators. A logistics operator should elaborate at least two scenarios and execution variants, as well as cause-and-effect analyses of successive project activities. All the above is well justified, since the analysis of time and flow of tasks requires thorough knowledge of the object of such activities, of the methods for executing them, and of the resources that determine their prompt completion; the determination of working conditions; the preparation of a project budget and simulations of expenditure; and the required cash flow for a given project. The analysis of different variants allows to elaborate flexible budgets and variant-based resource reservation, all of which serve to minimize risk. Planning future activities and how to use resources (own and third party), planning the conditions of execution of oversized cargo transportation, predicting conflicts and emergency situations, and preparing a coordination plan of logistics operations are all necessary for the effective execution of a given logistics project.

Resource planning results from expenditure planning. The type and nature of resources used for the purposes of carrying out a given logistics project depend on the nature, scope, and object of delivery. Due to its level of complexity, the execution of a particular logistics project requires access to diverse resources, both renewable and nonrenewable. The resources are often owned by third-party business entities; and for the purposes of carrying out a logistics project, they may be made available (temporarily) on certain terms as part of cooperation. The selection of business partners, i.e., the owners of resources, requires assessment and making a decision based on specified criteria. The criteria for assessing suppliers may be quantitative or qualitative. Due to the complexity of the process, it is proposed to apply appropriate methods that would support decision-makers. During resource planning, the selection of means of transportation suitable for a given cargo constitutes a significant issue. The basic parameters influencing selection are its size, structure, and weight. The number of axes, the loading capacity, the efficiency of the hydraulic system, and the power of the tractor unit are important parameters when selecting motor vehicles. Vehicles for transportation of oversized cargo are among the most technically advanced means of transportation.

The estimation of costs of oversized cargo transportation is a time-consuming process, and its level of uncertainty and risk is high. It is particularly difficult to assess the total cost in the planning phase due to the large number of variables and their nature. Each transportation service requires individual cost estimating. Each order is a specific one and therefore needs a special approach. The basic components of the cost of oversized cargo transportation include: loading and unloading, energy carriers (including fuel), drivers' remuneration, pilotage, route inspection, permits, expert evaluations, infrastructure adaptation, toll charges (including transportation on toll road sections), ferry crossings, cargo insurance, vehicle insurance, licenses, cargo security, tires, depreciation, order management, etc. Previous experience and appropriate methods for evaluating planned expenditure of performing the service are highly useful for assessing the total cost of oversized cargo transportation. Owing to previous experience, it is possible to accurately estimate the particular components of expenditure at each stage of a given logistics project. Dividing a particular logistics project (in this case, oversized cargo transportation) into separate components and assessing the expenditure on each component, followed by the aggregation of these estimates, can be helpful.

Based on observation of practice, logisticians typically leverage their knowledge, experience, and estimators to estimate logistics project costs, i.e., they usually rely on their intuition. Researchers have worked to develop cost estimators that maximize the practical value of limited information available in order to improve cost estimate accuracy and reliability, which should improve the suitability of resultant designs and subsequent project execution work. In practice there are many methods of estimation of costs of project accomplishment. Methods used to set project costs may be divided into three groups [20]:

- Approximation methods
  - Assessing by experts
  - Conscious guessing
  - Rough estimation
  - Empirical
  - By analogy
  - Parametrical methods
  - Cost estimation equations
  - Statistical
  - Probabilistic
  - Mathematical models
  - Detailed methods
  - On the basis of job batches
  - On the basis of work preparation methods
  - Other detailed methods

Using the abovementioned methods may help in achieving general estimates of project cost, but may fail with uncertainty. How does one perform estimation of uncertain project cost then? The total cost of a project is dependent on many variables, assumptions, and conditions. These variables influence significantly the size of the probable cost. The most difficult aspect of project cost estimating is gathering date of sufficient variety, quantity, and quality. The existing methods of cost estimation lead to determining quantity of total project cost as one single quantity, i.e., certain quantity as opposed to ambiguous quantity—uncertain. Single project cost may be treated as one of the possible options—a possible cost variant [21].

It is possible to employ methods based on fuzzy set theory into the practice of transportation-freight forwarding-logistics enterprises; such methods facilitate modeling the uncertainties characteristic of such undertakings. Triangular fuzzy sets are very often used in the practice because the parameters defining them can be easily specified in linguistic terms [22, 23]. Therefore, triangular fuzzy sets are applied to describe uncertain cost of activities [24]. Input parameter cost estimations are positive triangular fuzzy numbers (fuzzy cost of activities of the logistics project network). Output parameter, i.e., total logistics project cost, is also a triangular fuzzy number—interval (Fig. 4). The interval indicates that the project cost is estimated at this interval with α-confidence (α-cut). The α-confidence of fuzzy number of project cost $\tilde{K}$ is defined as follows [22]:

$$\tilde{K}_\alpha = \{x \in X : \mu_{\tilde{K}}(x) \geq \alpha\}, \quad \forall_{\alpha \in [0,1]} \tag{1}$$

Note that $\tilde{K}_\alpha$ is a crisp set rather than a fuzzy set. Using α-confidence, $\tilde{K}_\alpha$ can be represented by different levels of confidence intervals. Figure 4 shows the shape of fuzzy number with α-confidence level.

The membership function has the shape of a triangle (Fig. 4). The shape depends on parameters l, m, and u. Parameter l indicates an optimistic cost of an activity of a given logistics project, m most probable cost, and u pessimistic cost. In the point $x = m$, class t membership function assumes the value of 1, whereas in points $x = l$ and $x = u$, membership function assumes the value of 0. An uncertain logistics project cost can be described by a fuzzy number defined by its membership function. The cost of an activity has an uncertain character. The cost is "about 4.5." It means the cost of the activity can be in the range of 1–10.5 conventional cost units (c.c.u.). The optimistic cost of the activity is equal to 1 c.c.u., the most probable cost is equal to 4.5 c.c.u., and the pessimistic one is estimated as 10.5 c.c.u.

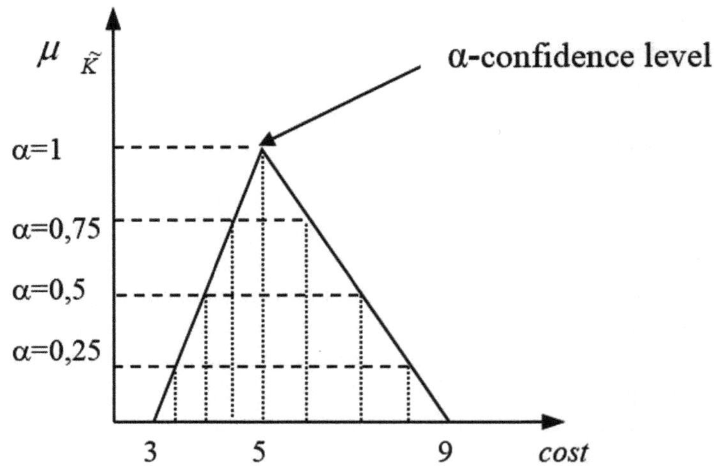

Fig. 4 The membership function of a given fuzzy cost $\tilde{T}_\alpha = (1; 4.5; 10)$ and its α-cut (source: own elaboration)

Table 4 Specification of a logistics project cost

Activity	Fuzzy cost [c.c.u.]	Fuzzy cost [c.c.u.]	Real cost [c.c.u.]
Cost of transport permit	"about 2000.00"	(1200.00; 2000.00; 2400.00)	1866.67
Cost of pilotage service	"about 1200.00"	(1000.00; 1200.00; 1600.00)	1266.67
Fuel cost	"about 1100.00"	(1000.00; 1100.00; 1400.00)	1166.67
Road taxes	"about 350.00"	(300.00; 350.00; 370.00)	340.00
Cost of vehicle driver	"about 550.00"	(500.00; 550.00; 700.00)	583.33
Cost of loading	"about 150.00"	(140.00; 150.00; 300.00)	196.67
Cost of unloading	"about 150.00"	(140.00; 150.00; 300.00)	196.67
Cost of load insurance	"about 900.00"	(800.00; 900.00; 1300.00)	1000.00
Total	"about 6400.00"	(5080.00; 6400.00; 8370.00)	6616.67

Source: own elaboration

In Table 4, cost assessments of an example of oversized cargo transportation are presented. The item of costs (activity cost of logistics project) is expressed in conventional cost units. The last column in the table concerns the real cost of a given project variant determined with the use of chosen method of defuzzification:

$$y = \frac{l + m + u}{3} \tag{2}$$

The fuzzy cost of the logistics project is in the range of 5080.00–8370.00 c.c.u. The optimistic cost is equal to 5080.00 c.c.u. The most probable cost is equal to 6400.00 c.c.u. and the pessimistic one is estimated as 8370.00 c.c.u. The single value of the cost of the logistics project realization is equal to 6616.67 c.c.u.

Oversized cargo transit constitutes a special type of transportation. It requires dedicated and thus costly equipment. It must be emphasized that in the case of mass transportation, a range of ready-made semitrailers can be readily purchased, whereas in the case of nonstandard transportation, these must be ordered beforehand (e.g., multi-axis modules are typically ordered approximately 2 years in advance). Entrepreneurs specializing in oversized cargo transportation are required to anticipate future market situation as well as the directions of economic development. For instance, a particular enterprise must assess whether or not it requires multi-axis semitrailers and to what kind of cargo will they be dedicated—low or high, long or extendable, perhaps heavy, etc. Entrepreneurs are expected to be able to suitably configure the vehicle and the semitrailer designed for the transportation of such cargos, since there is no single applicable standard. Such a configuration is particular to a given transportation order. Nonstandard cargo transportation is carried out with suitably fitted vehicles, the carrying capacity, structure, and labeling of which are usually different from those of standard means of transportation. In order to organize the transit of such vehicles, appropriate permits are required, as well as advance arrangements with appropriate infrastructure administrators. This involves obtaining specific permits and decisions of authorized entities.

Fig. 5 Example of special vehicle used in oversized cargo transportation (source: [26])

The transportation of oversized cargo requires special means of transportation as well as dedicated handling equipment (Fig. 5). Such cargo is characterized by large linear dimensions, the outline of which exceeds the dimensions of a vehicle or transportation unit. The nonstandard size, weight of the load, and pressure on the axis of the vehicle demand special care in the loading process. Securing oversized cargo on transport vehicles arises from the necessity to make it unmovable against the vehicle in such a way that it will arrive at the destination unchanged (geometrically, in shape and dimensions) and undamaged and at the same time not endangering people and the environment. This process frequently demands unconventional approach, also "oversized" due to the special parameters of the cargo. Securing such cargoes always demands individual calculations, assistance of the surveyor, and application of "unconventional" methods: welding, applying special fastening structures, cradles, etc. [11]. The process of securing oversized cargo can be aided by specific software system. One of the software systems is easyLOAD system developed by Goldhofer company. The program supports decision-making in the loading process. It is the first and only software on the market to be TÜV approved for transport engineering. The advantages of the easyLOAD system are the following [25]:

- Optimal planning with regard to the loading and configuration of the vehicle
- Documentation for their customers for professional transportation
- Clear loading instructions for the driver
- Facilitates transport authorization with the authorities
- Verification for transportation checks regarding the proper loading of the vehicle

An example of loading process planning is respectively shown in Figs. 6 and 7.

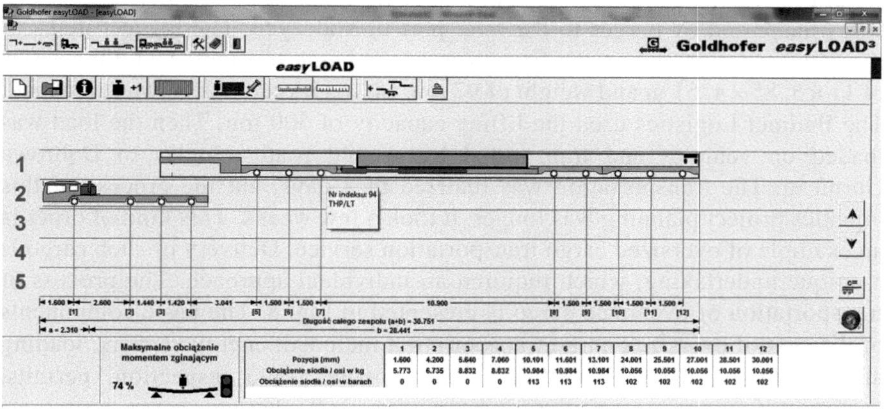

Fig. 6 An example of loading planning in easyLOAD system (source: own elaboration)

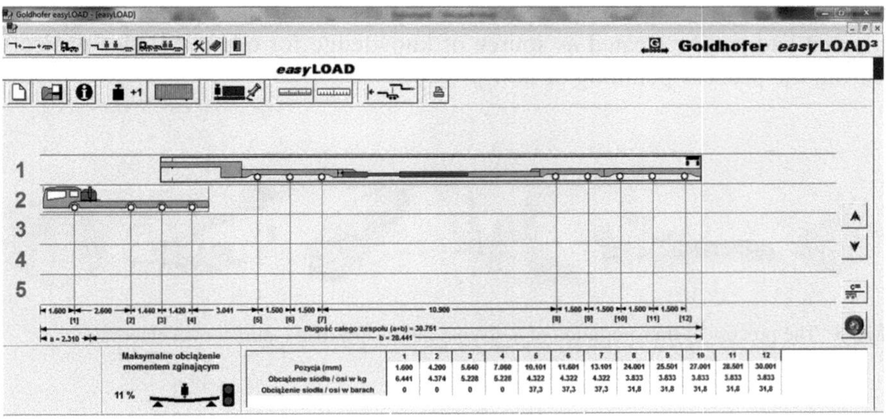

Fig. 7 An example of loading planning in easyLOAD system (source: own elaboration)

An Example of Transportation Service

We consider an example of oversized transportation realized by one of logistics providers. Bedmet Logistic Damian Bednarz Sp. K. is a Polish enterprise which specializes in heavy transport. The range of the enterprise operating includes: road transportation, inland waterway transportation, railway transportation, sea transportation, transshipment and storing, as well as relocations. A few months ago, the enterprise received an order to transport converter elements for Huta Katowice. The logistics project of transportation this kind of load needed special permits for road transportation. The enterprise was forced to do specialist expertise of several bridges, a roundabout reconstruction, the road marking dismantle, and dismantle and reassemble of about 100 light sirens. The oversized cargo sailed to the commercial port of Świnoujście in two parts. The total weight of the load equaled 800 ton. The route began in Świnoujście. The load was transported by barges to the river port in Malczyce. The biggest items of the converter elements in the first stage of the barge transport had dimensions of 11 × 5, 85 × 4, 51 m and weight of 97 ton. In Malczyce the load was unloaded. The Bedmet Logistics used the lifting capacity of 500 ton. Then the load was loaded on vehicles and transported by special road vehicles to Dąbrowa Górnicza. The transportation was realized in 3 days, but the process of this logistics project planning was longer. It took a few weeks. This kind of order is an example of oversized cargo transportation service. Delivery of such cargo is a unique undertaking, which required an individual approach. The process of transportation of oversized cargo is presented in Fig. 8. The basic components of the cost of oversized cargo transportation included: energy carriers, loading and unloading, drivers' remuneration, pilotage, route inspection, permits, expert evaluations, infrastructure adaptation, toll charges, barge crossings, cargo insurance, vehicle insurance, licenses, cargo security, tires, depreciation, order management, etc. This logistics project was realized in cooperation with other enterprises. The cooperation guaranteed the fulfillment of all of the order requirements. This new experience and the results of the project were documented and can be treated as source of knowledge for other logistics projects during the process planning of new projects.

Fig. 8 The process of transportation of a given nonstandard load (source: own elaboration)

Conclusions

The complexity of a logistics project in the context of oversized cargo transportation is mostly related to the object of the project, the sites where cargo is loaded and delivered to, the time frame of transportation, technical parameters of the cargo, its dimensions and weight, the number of distinct means of transportation used for delivering the cargo, their type, number and type of resources necessary for carrying out the project (including the means of loading and unloading), available infrastructure, the use of intermodal transportation, time of project execution, route of transportation, relationships with other orders, the number of subcontractors, the scope of outsourced work, legal regulations, elements of the project, and the relationships arising between different projects carried out by a given enterprise or a delivery/ supply chain. These factors constitute the basic sources of risk in the execution of a given logistics project.

The effective execution of oversized cargo transportation necessitates the use of suitable methods, tools, and techniques that make up together an appropriate approach to the management of such services. Our discussion is based on the statement that the execution of oversized cargo transportation exemplifies the fulfillment of short-term market needs. We draw an analogy between the management of orders for oversized cargo transportation services and project management. A given order can be treated as a project, and in this case, due to its specific features, as a logistics project. The planning of such a service corresponds to the planning of an undertaking characterized by uniqueness, thus involving nonroutine decision-making. We propose applying the project approach to the management of a given process—oversized cargo transportation, as part of carrying out a service.

References

1. Pisz, I., Łapuńka, I.: Conditions of logistics projects planning in transport-spedition-logistics sector on an example of oversize transportation. Part 1. Logistics, no. 3, 5134–5143 (2014) (in Polish)
2. Pisz, I., Łapuńka, I.: Conditions of logistics projects planning in transport-spedition-logistics sector on an example of oversize transportation. Part 2. Logistics, no. 3, 5143-5155 (2014) (in Polish)
3. Zhang, L.: Managing project changes: Case studies on stage interaction and functional interaction. International Journal of Project Management, no. 31, 958-970 (2013)
4. Vargo, S. L., Lusch, R. F.: Evolving to a new dominant logic for marketing. Journal of Marketing, no. 68, January, 1–17 (2004)
5. Vargo, S.L., Magliob, P.P., Akakaa, M.A.: On value and value co-creation: A service systems and service logic perspective. European Management Journal, vol. 26, iss. 3, June 145–152 (2008)
6. Stank, T.P., Goldsby, T.J., Vickery, S.K., Savickie, K.: Logistics service performance: Estimating its influence on market share. Journal of business logistics, vol. 24, no. 1, 27–55 (2003)

7. Soinio, J., Tanskanen, K., Finne, M.: How logistics service providers can develop value added services for SMEs: a dyadic perspective. The International Journal of Logistics Management, vol. 23, iss. 1, 31–49 (2012)
8. Pisz, I., Łapuńka I.: Project management in logistics. Difin, Warsaw (2015) (in Polish)
9. Bowersox, D.J., Closs D.J., Cooper M.B.: Supply Chain Logistics Management, McGraw-Hill/Irwin, New York, (2002)
10. Armstrong and Associates: Armstrong's Guide To Third Party Logistics Services Providers, vol. 7, Stoughton, WI: Armstrong and Associates, Inc, (1999)
11. Galor, W. (ed.): Carriage and securing of oversize cargo in transport. Maritime University of Szczecin, Szczecin, (2011) (in Polish)
12. Oversize Baltic. Streamlining: cross-border transportation management of oversize cargo. Baltic Transport Journal, no. 1, 30–39, (2014) (in Polish)
13. Jóźwiak, Z., Kawa, M.: The use of modern logistic solutions in transport of oversized cargo. Logistics, (2009)
14. O'Reilly, J.: Project Logistics: Moving the Big, the Heavy, and the Wide. Inbound Logistics, February, (2012), http://www.inboundlogistics.com/cms/article/project-logistics-moving-the-big-the-heavy-and-the-wide/
15. Krabbendam, R.: Are rules needed on trailer stability? Heavy Lift & Project Forwarding International. May/June, iss. 32, (2013)
16. Regulation of the Minister of Transport, Construction and Maritime Economy of 22 June 2012 on permits for oversized vehicles pass (Journal of Laws of 2012, item. 764), (2012) (in Polish)
17. Regulation of the Minister of Transport, Construction and Maritime Economy of 23 May 2012 on piloting oversized vehicles (Journal of Laws of 2012, item. 629) (in Polish)
18. ESTA information, http://www.esta-eu.org/
19. OSPTN information, http://ponadnormatywni.pl/
20. Madauss, B.J.: Handbuch Projektmanagement. Schäffer-Poeschel Verlang, Stuttgart, (1994)
21. Pisz, I., Banaszak, Z.: Project management subject to imprecise activity network and cost estimation constraints. [in:] Applied Computer Science, Świć A., Banaszak Z. (eds.), Lublin University of Technology, Institute of Technological Systems of Information, vol. 6, no 1, s. 7–28, (2010)
22. Zadeh L.A.: Fuzzy sets. Information and Control, vol. 8, 338-353 (1965)
23. Yoon, K.P.: A probabilistic approach to rank complex fuzzy numbers. Fuzzy Sets and Systems, no. 80, (1996)
24. Chen, S.-P.: Analysis of critical paths in a project network with fuzzy activity times. European Journal of Operational Research, no. 183, 442–459 (2007)
25. easyLOAD information, http://www.goldhofer.de/gh/pdf/en/News_10_04_e-k.pdf?WSESSIONID=58c0c29bc2e46df387c0486e76077c92
26. Bedmet Logistic information, www.bedmet.pl

The Impact of Traffic Parameter Assessment on Noise Emission

Justyna Sordyl

Abstract The study is an experimental attempt to determine the correlation between traffic parameters, meteorological parameters, and sound pollution (noise) created as a result of the traffic. This chapter presents research results of traffic noise level recorded in the street of the transit and distribution function, which is a part of the road system of the city of Bielsko-Biala, Poland. The study was conducted before and after the opening of the expressway S-69, which took over the transit function from the examined street. Measured noise data were compared to the registered traffic flows and the share of heavy vehicles in the traffic flow. The results of research conducted before and after opening of the expressway were also compared. The studies are at the first step towards obtaining general assumptions to the formulation of a semiempirical model of acoustic impact of traffic flow.

Keywords Noise emission • Sound emission • Equivalent sound level • L_{Aeq} • Traffic parameters • Traffic flow • Impact • Correlation

Introduction

Along with economic development there has been a continuous increase in the number of moving vehicles on Polish roads. This results in an increase in the level of impact of road transport on the acoustic climate communication routes through the issue of noise. In broad meaning, the noise is considered as undesirable, annoying, and causing fatigue sounds [1].

The methods against road traffic noise can be divided into two groups [2, 3]:

- Methods used “at the source”, reducing the formation of noise—creating vehicles having a lower sound power, using solutions which reduce emission (e.g. silencers), using modern road surfaces

J. Sordyl (✉)
Faculty of Management and Transport, University of Bielsko-Biała, Bielsko-Biała, Poland
e-mail: jsordyl@ath.eu

M. Bąk (ed.), *Transport Development Challenges in the Twenty-First Century*,
Springer Proceedings in Business and Economics,
DOI 10.1007/978-3-319-26848-4_14

- Methods reducing the spread of noise—improving acoustic insulation of buildings, construction of road tunnels, acoustic baffles, planning routes with consideration of the acoustic characteristics of ground, constructing ring roads

Due to high traffic load of the main routes in the city, in 2008 in Bielsko-Biala the construction of the north-eastern ring road started. It is a part of the S1 and S69 expressways and is designed not only to increase the capacity but also to reduce the degradation of roads in the city, by redirecting transit traffic in the north–south direction. The construction of the ring road made it possible to reduce the noise level in the city by about 3–5 dB [4].

The aim of the research done by the author and presented in this chapter was to evaluate the impact of traffic parameters on the level of traffic noise in Żywiecka Street. This street was chosen because it has some of the highest levels of sound in the city [5]. This street has a transit and distribution function in the traffic system of Bielsko-Biala. It is a two-way street of the fourth technical grade with two lanes. A part of the research was conducted after the opening of the road S69, which took over a part of the traffic flow. The S69 is an expressway and it is a part of Trans-European Transport Network. It is also a part of the sixth Pan-European Corridor and a part of a TEN-T priority project No. 25 [6]. The location of Żywiecka Street and of the measuring point in the city road system is shown in Fig. 1.

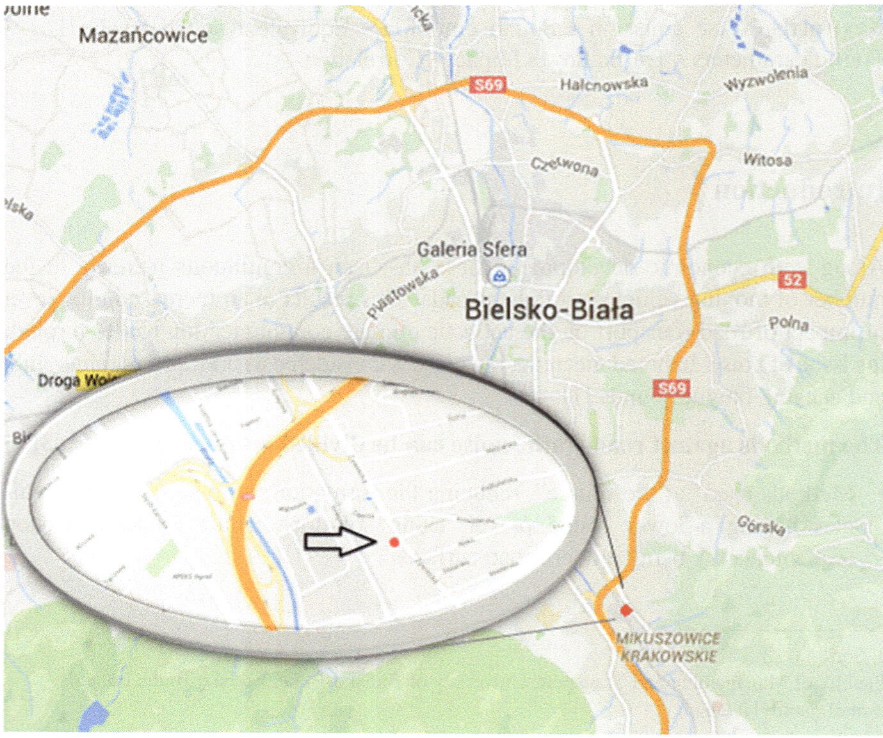

Fig. 1 Location of Żywiecka Street and the measuring point in the city road system (source: Google maps)

The studies are at the first step towards obtaining general assumptions related to the work, which is to lead to the formulation of a semiempirical model of traffic flow acoustic impact.

Research Methodology

The noise level tests in the selected location of Żywiecka Street were conducted using a set of measuring equipment adapted for use in different weather conditions. The measurement set installed at the measuring point as in Fig. 2 includes the following devices:

- Stand and microphone with all-weather cover
- Integrating Sound Level Meter SON-50
- Weather Station Lab-el
- GoPro Hero3 digital camera

During the tests, the microphone was placed on a special stand equipped with a microphone extension cable. It allowed to make tests at a height of 4 m above the ground and a distance of 4 m from the edge of an external traffic lane. Before the tests, a calibration of sound level meter was made. Weather Station Lab-el (with as visible in Fig. 2 wind speed and direction sensor) is a set of sensors for continuous measurement of temperature, humidity, atmospheric pressure, and

Fig. 2 The measurement set installed at the measuring point (source: own materials)

wind parameters (speed and direction). The use of the weather station allowed for the current assessment of local meteorological conditions that determine the admissibility of measurements [7]. Permissible limiting values are set out in the Ordinance of the Minister of Environment [8] and are shown later in the paper. During the tests, thanks to a digital camera, the intensity and the generic structure of the passing vehicles were recorded. The main specification of the measuring equipment is presented in Table 1.

Noise level measurements were made in two basic stages. The first stage was conducted when the tested street had a transit function for the city. The second stage was conducted after the opening of the S69, which took over part of the traffic load. To accomplish the research task, two working days were selected (one before the opening of the expressway, the second after) during which continuous measurements of equivalent sound level between the hours of 5 pm and 7 pm were conducted.

Table 1 The main specification of the measuring equipment (source: devices technical specification sheets)

Sonometer SON-50	
Accuracy class	1
Measurement microphone	½″
The range of operating temperature	−10 to +50 °C
The range of relative humidity	≤90 % (non-condensing)
The range of atmospheric pressure	65–108 kPa
Corrective frequency characteristics	A, C, LIN
Dynamics characteristics	SLOW, FAST
The total measurement range of equivalent sound level (L_{Aeq})	20–135 dB
Acoustic calibrator KA-50	
Accuracy class	1
Nominal level of sound pressure	94.0 dB
Nominal frequency	1000 Hz
Weather station LB-755A	
The range of temperature measurement	−40 to +85 °C
The range of humidity measurement	10–95 %
The range of pressure measurement	700–1100 hPa
The range of wind speed measurement	1–100 m/s
The range of anemometer rotation	360°
GoPro Hero3 digital camera	
Lens	f/2,8, wide angle with reduced distortion
Video resolution	From 4 K (up to 15 fps) to WVGA (up to 240 fps)

At the same time, along with the noise levels, there were recorded:

- Traffic volume and the generic structure of vehicles forming a traffic flow: It has been obtained thanks to the use of a digital camera.
- Meteorological parameters (wind speed and direction, atmospheric pressure, temperature, and humidity): It has been obtained thanks to the use of a weather station.

The sound level record was made with the frequency of 1 Hz and then an equivalent sound level with background noise was calculated.

As the continuous measurements were performed with periodic registration of $L_{\mathrm{Aeq\ tk}}$ sound level, the calculation of equivalent sound level with acoustic background L_{Aeq0T} for 1.5-h time reference covering a wide range of periods of registration requires to use the following conversion [7]:

$$L_{\mathrm{Aeq0\ T}} = 10\log\left[\frac{1}{T}\sum_{k=1}^{n} t_k 10^{0,1L_{\mathrm{Aeqt_k}}}\right] \tag{1}$$

where: T—reference time interval [s],

t_k—the period of registrating results [s],

$L_{\mathrm{Aeq\ tk}}$—equivalent sound level A-weighting during the period of recording results t_k in decibels [dB],

n—the number of result registration periods t_k.

The next step is to determine the reference noise generated by different sources than road vehicles whose impact cannot be eliminated at the time of measurement. In this chapter for the reference sound level, the minimum value of the measured noise level was assumed. They occurred during the corresponding 1.5-h reference time intervals.

Determination of equivalent sound level A-weighting for the reference time interval requires to use the following conversion [7]:

$$L_{Aeq\ T} = 10\log\left(10^{0,1L_{Aeq0T}} - 10^{0,1L_{Aeq\,\mathrm{Ref}}}\right) \tag{2}$$

where: $L_{\mathrm{Aeq\ Ref}}$—the reference sound level in decibels [dB].

Test Results

The times of occurrence of particular noise level classes in percentage terms (Fig. 3) show that in the second stage of the study recorded sound levels were much lower than in the first stage. The results indicate that, prior to the opening of the expressway, the noise levels exceeding 65 dB (– max value permitted in Polish law regulations [8]) were recorded at 70 % of the measurement time, and after the opening only at 34 %.

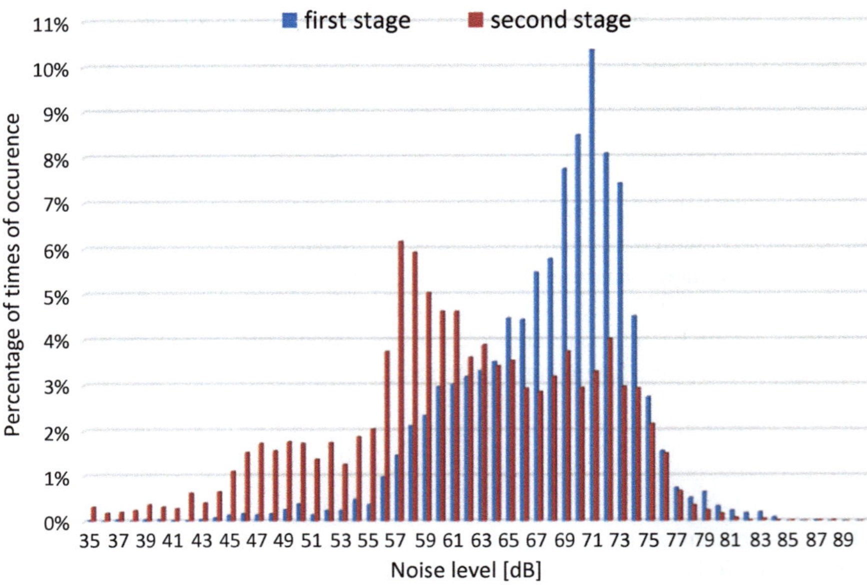

Fig. 3 Times of occurrence of particular noise level classes in percentage terms recorded in the first and the second stages of the research (source: own materials)

Table 2 Comparison of equivalent sound level $L_{Aeq1,5h}$ with traffic flow (source: own materials)

	Stage 1	Stage 2
Equivalent sound level $L_{Aeq1,5h}$ [dB]	70.4	67.8
The average number of vehicles per hour	1091.0	388.0
The average number of passenger vehicles, passenger—delivery and motorcycles per hour	998.0	368.0
The average number of trucks, buses, other special vehicles per hour	93.0	20.0

Table 2 combines equivalent sound levels L_{Aeq} calculated for the 1.5-h time period with the total number of vehicles and the number of vehicles divided into two basic categories.

Figure 4 compares the average noise level with reference to the traffic flow, with division to the generic structure of the vehicles. It is easy to notice that in the second stage of the research both recorded sound levels and traffic congestion were significantly lower than in the first stage.

The comparison of the calculated equivalent noise level with reference to the registered traffic flow for each test minute is shown in Fig. 5. The value of the correlation coefficient (Table 3) indicates the existence of moderate positive interdependence of noise level on traffic flow. With a small variation of traffic

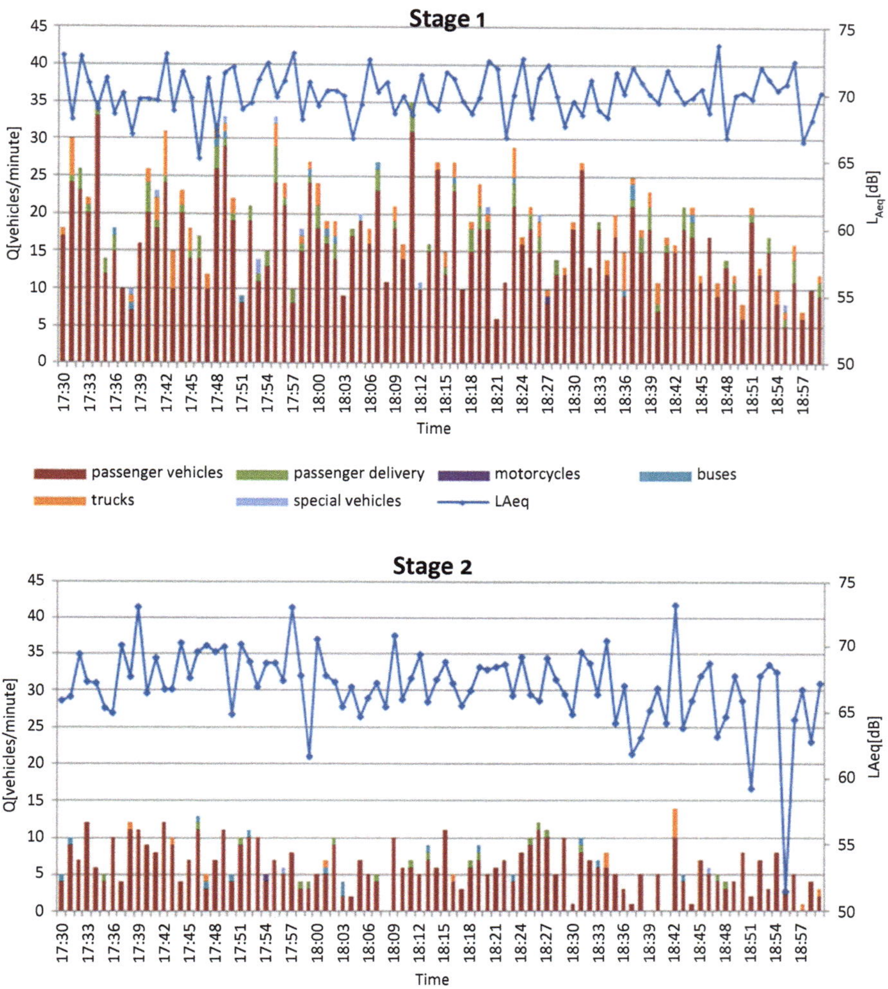

Fig. 4 Comparison of the traffic flow to the average noise levels (source: own materials)

volumes and limited number of tests it is difficult to define a clear dependence of noise level on traffic flow. It is also a result of the change of generic structure, related to the change in the share of heavy vehicles. The resulting dependence of the calculated equivalent noise level with reference to share of heavy vehicles (calculated as the sum of trucks, buses, and other special vehicles) for each test minute is shown in Fig. 6. As it might be expected, with comparable traffic volume, the noise level is largely determined by the share of heavy vehicles in the traffic structure (Table 3).

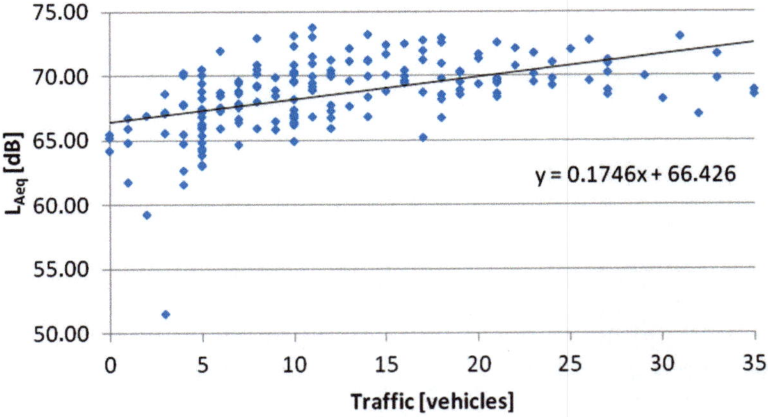

Fig. 5 Comparison of equivalent sound level L_{AeqT} to traffic flow with defined trend line (source: own materials)

Table 3 Correlation of recorded data (source: own materials)

	L_{Aeq}	Traffic flow	Share of heavy vehicles
L_{Aeq}	1.00		
Traffic flow	0.49	1.00	
Share of heavy vehicles	0.24	0.18	1.00

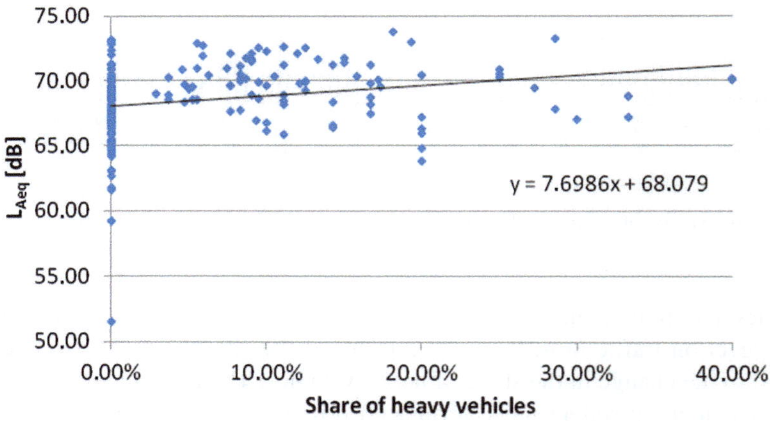

Fig. 6 The dependence of the calculated equivalent noise level with reference to share of heavy vehicles with defined trend line (source: own materials)

Conclusion

The analysis of recorded research material indicates that the opening of the S69 had a significant impact on improving the acoustic climate in Żywiecka Street. Equivalent sound level $L_{Aeq1,5}$ declined by 2.6 dB, and recorded traffic was reduced by 35 % and the number of heavy vehicles by 21 %. The new expressway took over the transit function from the examined street and had huge impact on recorded traffic flow.

The acceptable level of noise in the downtown area for cities with more than 100,000 residents for the day L_{AeqD} (counted from 6 am to 10 pm) is 65 dB [8]. Knowing that the research was done beyond the rush hours of the tested road [9], it cannot be excluded that despite the reduction of the noise level, exceeding acceptable noise levels may still occur.

With a small variation of traffic volumes and limited number of tests, it is difficult to define a clear dependence of the noise level on traffic flow but the value of the correlation coefficient (0,49) indicates existence of correlation. When the traffic volume is comparable then the noise level is determined also by the share of heavy vehicles in the traffic structure.

It has been noted that the impact of meteorological parameters on the recorded sound levels is not defined and difficult to determine with such small amount of collected data. However, the use of the weather station enabled immediate evaluation of the local weather measurement conditions. The acceptable limit values of meteorological parameters are defined in the Ordinance [7] and are shown in Table 4 along with parameters recorded during the test. There have been no exceedances of the limit of meteorological parameters.

Table 4 The conditions under which the measurements may be conducted together with recorded meteorological parameters

Parameter name	The limit values	The values recorded
Temperature	−10 to 50 °C	5.4–13.2 °C
Relative humidity	25–98 %	60–75 %
Wind speed	0–5 m/s	0–4.8 m/s
Atmospheric pressure	900–1100 hPa	950–961.4 hPa
Precipitation	Lack	Lack

Source: [7], own materials

References

1. Kucharski, R.J., Taras, A., Chyla, A., Szymański, Z., Kraszewski, M. (red.): Zanieczyszczenie Środowiska hałasem w Świetle badań WIOŚ w 1999 roku. Inspekcja Ochrony Środowiska, Warszawa (2000) [Kucharski, R.J., Taras, A., Chyla, A., Szymański, Z., Kraszewski, M. (eds.): Noise pollution of the environment according to the survey of the Regional Inspectorate of Environmental Protection in 1999. Inspection for Environmental Protection, Warsaw (2000)]
2. Gronowicz J.: Ochrona Środowiska w transporcie lądowym. Instytut Technologii Eksploatacji, Radom (2004) [Gronowicz J.: Environmental protection in land transport. Institute for Sustainable Technologies, Radom (2004)]
3. Raport o stanie Środowiska 2008. Główny Inspektorat Ochrony Środowiska, Biblioteka Monitoringu Środowiska, Warszawa (2010) [State of environment report 2008. Chief Insectorate of environmental protection, Environmental Monitoring Library, Warsaw (2010)]
4. Sordyl, J.: Badania poziomu hałasu komunikacyjnego w warunkach szczytu komunikacyjnego. Technika Transportu Szynowego, 2013, No. 10, 2969–2975 [Sordyl, J.: Studies of traffic noise level in the rush hours. Railway Transportation Technology, 2013, No. 10, 2969–2975]
5. Program Ochrony srodowiska przed hałasem dla miasta Bielska-Białej, Ekkom Sp. z o.o., Kraków (2013) [Environmental protection program against the noise of Bielsko-Biala city, Ekkom Ltd., Cracow (2013)]
6. Umowa na kontynuację budowy S69 podpisana, Generalna Dyrekcja Dróg Krajowych i Autostrad [The agreement on the continuation of the construction of S69 signed, General Directorate for National Roads and Motorways]
7. Rozporządzenie Ministra Środowiska z dnia 16 czerwca 2011 r. w sprawie wymagań w zakresie prowadzenia pomiarów poziomów substancji lub energii w Środowisku przez zarządzającego drogą, linią kolejową, linią tramwajową, lotniskiem lub portem Dz. U. 2011 Nr 140 poz. 824 [Ordinance of the Minister of Environment of June 16, 2011 on the requirements for the measurements of substances or energy levels in the environment by road, railway, tram line, airport or port manager. Journal of Laws of 2011, No. 140, item 824]
8. Rozporządzenie Ministra Środowiska z dnia 1 października 2012 r. zmieniające rozporządzenie w sprawie dopuszczalnych poziomów hałasu w Środowisku. Dz.U. 2012 Nr 0 poz. 1109 [Ordinance of the Minister of Environment of October 1, 2012 amending the ordinance on the permissible noise levels in the environment. Journal of Laws of 2012, No. 0, item 1109]
9. Stanclik, M., Brzozowska, L,: Poziom hałasu a struktura i natęŻenie ruchu pojazdów – analiza dla wybranej drogi w Bielsku-Białej. W: Maczyński, A. (red): Wspomganie komputerowe i logistyka w transporcie. Wydawnictwo Naukowe Akademii Techniczno-Humanistycznej, Bielsko-Biała (2014) [Stanclik, M., Brzozowska, L,: Noise level vs vehicle structure and traffic flow - analysis of selected road in Bielsko-Biala. In: Maczyński, A. (ed): Computer support and logistics in transport. University of Bielsko-Biala Scientific Publisher, Bielsko-Biala (2014)]

Identification of Behavioral Changes in Transport as a Means for Decarbonisation of the Economy

Urszula Motowidlak

Abstract In the Transport White Paper, the European Commission delivers its transport strategy up to the year 2050. One of the key objectives is defined as the 60 % reduction of greenhouse gas emissions (GHG) in transport by 2050. In order to achieve this objective, a number of decisive actions need to be undertaken. In comparison to other sectors, the GHG emissions in transport are predicted to increase continuously. The research has shown that behavioural changes may enable considerable energy savings and foster substantial reduction of CO_2 emission.

Keywords Transport and energy consumption • CO_2 emission • Drivers' behavior • Eco-driving

Introduction

The purpose of this chapter is to analyse changes in driver's behaviour and identify courses of action that may encourage users to select eco-friendly forms of transport.

Mobility is the foundation of modern economy. Private and public transport guarantees greater flexibility in everyday life. A well-functioning transport sector enables companies to create sophisticated supply chains and to provide sufficient and effective delivery of goods and transportation of people. All this aims to improve the functioning of markets for goods and labour and contributes substantially to GDP. Simultaneous development of the automotive industry increases greenhouse gas emissions generated by the transport sector. Transport is responsible for approximately a quarter of the total GHG emissions within the European Union (EU). The road transport is responsible for over two-thirds of the transport emissions. Unlike in many other sectors, the GHG emissions of transport are predicted to continue to increase. One of the key objectives set in the White Paper is the 60 % reduction of transport GHG emissions

U. Motowidlak (✉)
Faculty of Economics and Sociology, Department of Logistics,
University of Lodz, Lodz, Poland
e-mail: umotowidlak@onet.eu

M. Bąk (ed.), *Transport Development Challenges in the Twenty-First Century*,
Springer Proceedings in Business and Economics,
DOI 10.1007/978-3-319-26848-4_15

by 2050 [1]. Changing people's transport behaviour may contribute to achieving this objective. One may identify a number of behavioural changes in transport that could contribute to the decarbonisation of transport. The present scholarship distinguishes four important aspects of such behaviour change, namely:

- Buying and using an electric car or a plug-in hybrid
- Buying and using a smaller car
- Applying a fuel-efficient driving style
- Making use of information and communication technology (ICT) to decrease business travel: teleworking and applying virtual meetings

For these four behavioural changes detailed assessments were carried out to estimate their potential for reduction.

The Directions of Changes in Energy Consumption and CO_2 Emissions in Transport in the EU

Globally, transport is the sector with the highest final energy consumption and, without any significant policy changes, it is likely to remain so. The sector is responsible for about 20 % of CO_2 emissions in the EU, and emissions from the sector are still rising rapidly [2].

It is widely recognised that the transport sector remains one of the most challenging areas for improving energy efficiency, and that while in the past measures focused on technological improvements, there appears the need to grasp a more holistic approach that would include reduction of transport demand and shift of transport to more environmentally friendly and energy-efficient modes. This is well summarised in Table 1:

Table 1 Sustainable hierarchy for transport measures

Activity	Target	Characteristic
Avoid	Reduce demand for powered transport	Wide range of measures ranging from good spatial planning through to technological solutions such as telecommuting
Shift	Modal shift to more sustainable modes	Shifting away from motorised modes to cycling and walking Shifting from private motor vehicles to public transport Includes better integration between different public transport systems, walking and cycling
Improve	Efficiency improvements	Behavioural changes including encouraging higher occupancy rates for both private vehicles (e.g. lift sharing) and public transport; promotion of car clubs; promotion of eco-driving techniques Technical interventions to improve vehicle efficiency Technical interventions to promote more efficient use of transport infrastructure and networks

Source: [15, p. 8]

Historically, the tendency to focus on technological solutions to improve the efficiency of vehicles, in particular of cars, dominated. A review of the energy efficiency action plans suggested that although there have been significant improvements over recent years in vehicle technology, particularly fuel efficiency, these were not sufficient to neutralize the effect of increases in traffic and car size. The EU, therefore, decided to tackle transport sector energy consumption in a more comprehensive way. The major challenge facing the transport system is the reduction of its fossil fuel consumption and carbon footprint [3].

Energy Consumption

Transport has shown the most rapid energy consumption growth out of all sectors. In consequence, its share of total final energy consumption increased from 26 % in 1990 to 32 % in 2012 (Fig. 1). Road transport accounted for 82 % (range of 64–97 %) of total EU transport consumption, and of this about 60 % was absorbed by cars. Trucks and light commercial vehicles consumed 30 % of the total amount of fuel in the transport sector in 2012. Respectively, air transport accounted for 13 % of the energy consumption in transport, while rail and water represented only 2 % each.

The increase in energy consumption in the transport has largely been driven by the growth in passenger mobility (Fig. 2), particularly the use of cars. Cars require up to 300 % more energy per passenger km (pkm) than public transport (trains, metro, tramways and buses), and 600 % more than rail transport. The level of mobility varies significantly between EU countries due to differences in car ownership levels, income, country size and density, ranging from 4200 pkm per capita in Romania to over 16,000 pkm per capita in Luxembourg. In general, mobility has increased most

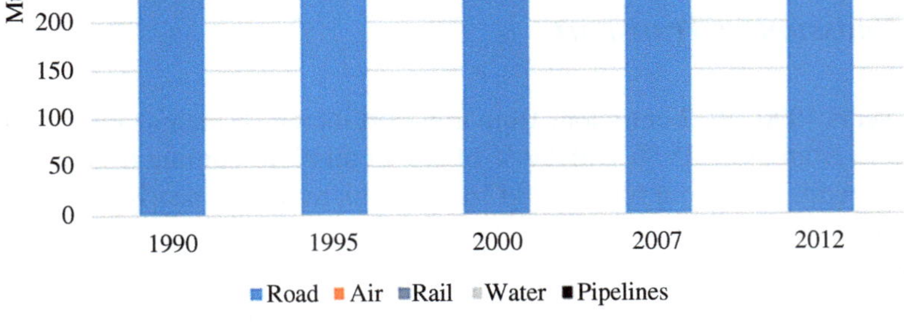

Fig. 1 Energy consumption by mode in EU (source: based on data of ODYSSEE MURE)

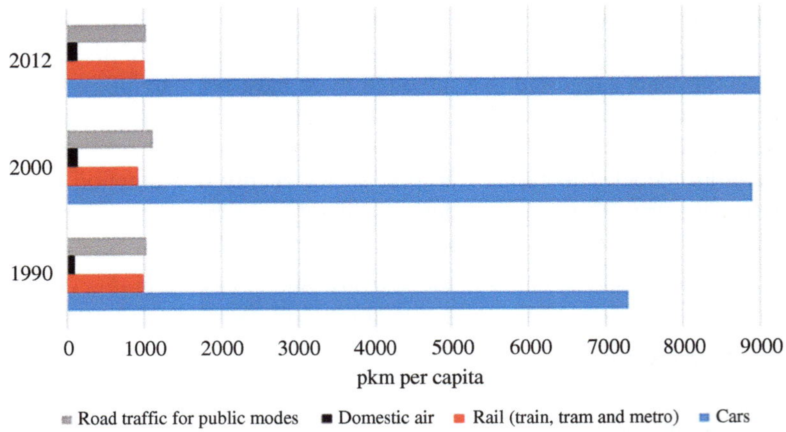

Fig. 2 Average annual passenger km per capita in EU (source: based on data of ODYSSEE MURE)

significantly in majority of new member states (the increase of over 5 % per year in Bulgaria, Latvia, Lithuania, Poland) as well as Greece and Ireland [4].

The bulk of the consumption in the sector is made up of oil products (the average of 95 % in the EU in 2012). Alternative fuels (electricity, natural gas and biofuels) supplied around 5.5 % of the consumption of road transport in the EU, of which 92 % were biofuels. Such a good performance is explained mainly by the rapid penetration of biofuels, and especially of biodiesel, following the prompt implementation of the EU Directive on biofuels.

Five countries have a high share of alternative fuels: Italy, Germany, Austria, Sweden and Spain (between 6.5 and 8.2 %). In 2012 Spain and Sweden were most efficient in this aspect (approximately 8 % alternative fuels). Italy and Bulgaria had the highest penetration of CNG (over 2 %). Electricity supplied less than 2 % of the consumption of transport in the EU in 2012. The highest share was observed in the Czech Republic, Austria and Sweden (approximately 3 %).

Development of road transport is not only responsible for the increase in demand for energy resources, but also for the growth of CO_2 emissions.

CO_2 Emission in Transport

In the years 1990–2012 emissions from transport increased their share of total CO_2 emissions from 19 to 25 %, with 90 % of these emissions coming from road transport. The emissions from road freight transport increased by nearly 36 % between 1990 and 2012 and made over 35 % of the sector's emissions (compared to 31 % in 1990). Emissions generated by cars increased by 15 %. At the same time, emissions generated by domestic air transportation vehicles increased by 17 %. They represent, however, less than 3 % of the total amount (Fig. 3).

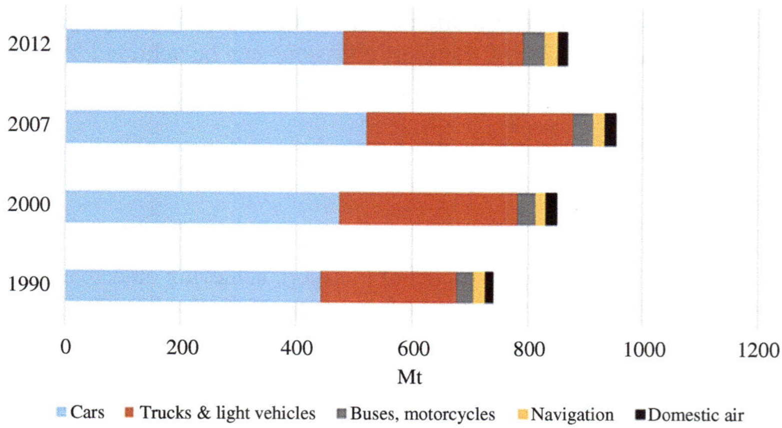

Fig. 3 CO_2 emissions from transport in EU (source: based on data of ODYSSEE MURE)

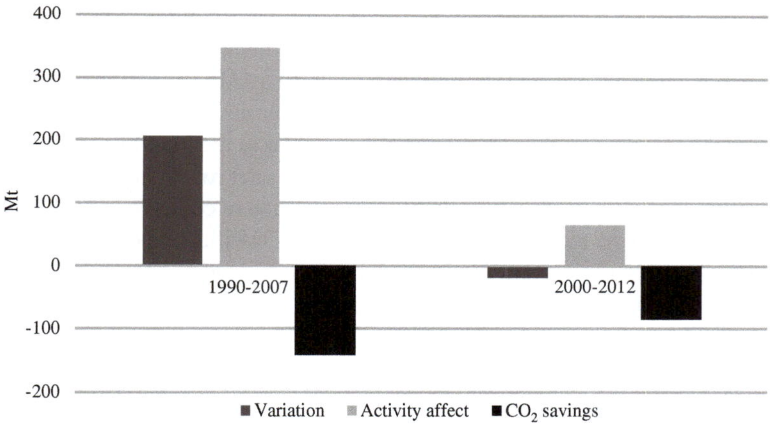

Fig. 4 Variation of CO_2 emissions from transport in EU (source: based on data of ODYSSEE MURE)

CO_2 emissions of transport have slightly decreased since 2000 (around −2 %). All modes participated to that reduction, except trucks and light vehicles (+2 %), which achieved 2 % increase in CO_2 emissions.

More than half of the increase in CO_2 emissions between 1990 and 2007 was offset by CO_2 savings [5]. These savings limited the increase in CO_2 emissions to 208 Mt (Fig. 4). Around 40 % of the savings come from trucks and light vehicles and 30 % from cars. The savings of CO_2 emissions amounted up to around 86 Mt in 2012.

The deployment of new clean fuels and transport technologies would be necessary to maintain the current service level and reduce emissions. The data presented above shows that the implementation of the new clean fuels and transport technologies will be necessary to maintain the projected increase in transport while reducing emissions. Growth in demand for petroleum resources while increasing market share of diesel cars indicates the need to undertake active measures in terms of creating a competitive, if resource-efficient, transport system [6]. High potential for energy savings and reduction of transport emissions can be achieved by changing certain modes of behaviour of drivers.

Driver Behaviour

The transformation of the current model of transport, achieving objectives that were defined in the White Book, and concerns reducing the dependence on fossil fuels and reducing CO_2 emissions, cannot be obtained simply by technical solutions. All this requires formulation of a new concept of mobility. E. Załoga noticed that many-sided analysis and interdisciplinary studies are needed to detect modern-day problems of transport process. Similarly to the above-mentioned studies, the ones suggested by Załoga should contribute to the new quality of transport. Many-sided attitude to the information about efficient energy use in the transport's sector has been proven by scholarship. The new quality of low-emission transport should be based on technological innovation, organisational rationalisation and behavioural changes.

The behavioural changes in transport have been regarded as a significant instrument for years. Thanks to this tool there is a possibility that the already existing strategies can consume less energy in a short-term perspective. As far as the long-term perspective is concerned, the behavioural changes are regarded as a vital role in ecological transport and logistics. Such goals were set by the stronger position of ecological mobility values and wider access to the technological improvements. Classification of these changes in transport's decarbonisation process is a relatively new studies' course. The lack of comprehensive studies on this matter is striking in scholarship. Such studies would certainly delineate the impact of these changes on effective management of energy in transport and their consequences on economy and society. According to the report conducted by CE Delft there are significant changes in the transport's behaviour. These changes prove to have a high potential in emission's reduction.

Buying and Using an Electric Car, Plug-In Hybrid and Smaller Cars

Authors of the report are convinced that electric cars guarantee most effective profits for the environment since when driving they emit no pollution whatsoever. Table 2 suggests that the CO_2 emission level can be reduced up to 90 % per passenger provided that we use electric cars. In the case of plug-in hybrid cars this level can be reduced to the maximum 69 % per pkm. The level of CO_2 emission depends also on the way the electricity is generated. For this reason there will be one level of CO_2 reduction in Poland, where the electricity is mainly generated from hard and brown coal, and another in France, where 50 % of electricity comes from nuclear power or in Germany, where the renewable sources generate around 30 % of energy.

Table 2 Maximum realistic CO_2 mitigation potential of buying and using an electric or plug-in hybrid car

Behavioural change	2020	2030	2050
Buying and using electric cars:			
Relative reduction in CO_2 emissions per pkm	19–34 %	64–72 %	82–90 %
Absolute CO_2 mitigation potential (Mton)	96–174	330–371	420–462
Buying and using plug-in hybrid cars:			
Relative reduction in CO_2 emissions per pkm	11–22 %	39–56 %	49–69 %
Absolute CO_2 mitigation potential (Mton)	56–113	198–286	251–354

Source: [14, p. 25]

When analysing the potential benefits of electric vehicles as the power source, there appear several significant advantages. Firstly, it makes the user independent from the supply and prices of oil fuels. It also eliminates the gearbox in the car and systems responsible for distributing the oil in the engine. Furthermore, the electric motor is much more effective than internal combustion [7]. The operating costs are seen as another advantage of such vehicles—100 km ride should cost several pennies, depending on the electricity tariff. Moreover, the owners of electric cars can also expect lower costs of servicing the car and liability insurance [8]. On the other hand, this type of a car is characterised by a limited battery duration so that it needs to be replaced every few years. What is more, the battery has got limited capacity. Besides, there is no extensive network of vehicle charging as yet, and the user has to be prepared for higher purchase costs [9].

The characteristics of development of the electric vehicle market in the EU indicate that interest in these vehicles is gradually increasing. The survey, which was carried out for the needs of the international project eMAP, states that in 2012 in Germany there were more than 7100 electric cars, in Finland 109 units, while in Poland 70 pieces [10]. However, the number of plug-in hybrid cars amounted to 65,000 in Germany and 6100 in Finland (figures for the Polish market are unknown). However, according to The European Automobile Manufacturers' Association (ACEA) the demand for new electric cars in the first quarter of 2015 increased by almost 120 % compared to the first quarter of 2014. The total number of registered electric cars reached therefore the level of 24,630 cars. The demand for new hybrid cars also increased in this period (21.4 %), reaching the total number of 56,704 units [11].

Limited popularity of electric cars in Poland is confirmed by the data provided by Automotive Market Research Institute SAMAR, which indicates that in December 2014 only four electric cars were registered in Poland. Yet, in 2013, within the same period of time, the number of registered electric cars was nine. Today, the number of cars and buses powered by electricity amounts in Poland to 1007 units [12].

Taking all this into consideration, one may conclude that in short-term perspective electric vehicles and plug-in hybrid cars, despite their positive impact on the process of decarbonisation of transport, will moderately contribute to the discussed process. Prices of electric cars and their performance are still the most important

Table 3 Maximum realistic CO_2 mitigation potential of buying and using smaller cars

Behavioural change	2020	2030	2050
Relative reduction in CO_2 emissions per pkm	17–20 %	18–21 %	18–21 %
Absolute CO_2 mitigation potential (Mton)	80–96	74–88	71–84

Source: [14, p. 32]

factors in the purchasing process. Because of their high prices, electric vehicles should be introduced to public transport first. Furthermore, there should be a possibility to rent an electric car. As instruments supporting measures for the dissemination of ecological electric car promotions actions can be pointed out. Energy savings can be achieved through the purchase and use of smaller cars (Table 3).

The figures presented in Table 3 suggest that smaller vehicles (850 kg) can reduce CO_2 emission up to 20 % as they are characterised by lesser fuel consumption. Meanwhile, when analysing the behaviour scenarios in the automotive market, we can see a stable tendency to purchase heavier and more powerful cars. In the period of 1995–2007, the power of newly bought cars increased from 64 to 96 kW in the EU 15. Improvement of fuel efficiency and achieved energy savings were very largely consumed by the new trends in the automotive industry. Since 2008, we observe a gradual increase in interest in smaller cars in the EU. According to data from the International Energy Agency (EEA), the average power of purchased cars decreased by 14 % and amounted to an average of 83 kW [13].

Eco-Driving

The fuel-efficient driving style includes a number of techniques [14]:

- Ensuring that the engine is used in the most efficient way, for instance, by early gear changes and less rapid braking and acceleration
- Applying the rules of minimising redundant energy use, i.e. by avoiding unnecessary air-conditioning, removal of unused roof racks and weight, minimising idling
- Maintaining tyre pressure at the appropriate level

Implemented in 2010–2013 EcoWill project, which was financed with EU funds, aimed at increasing the awareness of drivers about the benefits of eco-driving. Drivers using eco-driving techniques can have a real impact on reducing CO_2 emission in road transport. Since 2008, eco-driving training in Austria has become a mandatory requirement for obtaining a driving license. It applies to drivers of passenger cars, trucks and buses [15].

The estimates for the maximum realistic mitigation potential of applying a fuel-efficient driving style are shown in Table 4.

Table 4 suggests that the eco-driving may result in the 10 % reduction of CO_2 emission by 2020. In the medium- and long-term perspective, CO_2 reduction level will probably decrease because of further improvements in the fuel efficiency.

Table 4 Maximum realistic CO_2 mitigation potential of applying a fuel-efficient driving style

Behavioural change	2020	2030	2050
Relative reduction in CO_2 emissions per pkm	10 %	7 %	2 %
Absolute CO_2 mitigation potential (Mton)	47	32	10

Source: [14, p. 34]

Table 5 Potential share of passenger kilometres and CO_2 emissions to be saved by applying teleworking and virtual meetings

Behavioural change	2020	2030	2050
Commuting trips:			
Relative share pkm commuting in total number of pkm of passenger transport (%)	13	12	12
Relative share CO_2 emissions due to commuting in total CO_2 emissions of passenger transport (%)	11	10	9
Business trips:			
Relative share pkm business trips in total number of pkm of passenger transport (%)	11	12	12
Relative share of CO_2 emissions due to business trips in total CO_2 emissions of passenger transport (%)	11	13	13

Source: [14, p. 37]

Teleworking and Applying Virtual Meetings

The development of communication leads to the increasing number of motor vehicles. This fosters interference and congestion in the road network. In consequence, the traffic flow in the city is reduced. Congestion is most evident in the mornings and afternoons. This is caused mainly by the commuters. In Moscow during the morning rush traffic congestion the rate amounts up to 111 %, while in the afternoon it is still higher and oscillates at 141 %. For Warsaw this level reaches, respectively, 71 % and 75 % in the morning and evening peaks. This means that if an average commuter spends 30 min on travel in the morning, commuting time is increased by more than 20 min in the evening. It has been calculated that the typical inhabitant of Warsaw spends approximately 12 days in a traffic jam per year [16]. Increased city traffic and traffic problems are prone to generate extensive environmental pollution. When driving slowly in heavy traffic, one increases emission of harmful substances, such as nitric oxide, sulphur dioxide, carbon monoxide as well as greenhouse gases [17]. Therefore, the change of working style and the usage of new techniques for virtual work can contribute up to 13 % of reduction in CO_2 emissions (Table 5).

The theoretical maximum emission reduction potential of both teleworking and applying virtual meetings is determined by several factors. The most important ones are:

- The number of vehicle kilometers substituted by teleworking and/or visual meetings
- Reduced energy use in the office (only for teleworking)
- Increased energy use at home (only for teleworking)
- Energy use of the ICT appliances used [14]

Conclusion

The review of behavioural changes indicates that there is significant potential for reducing fuel consumption and its emission. It should also be emphasised that the possibility of using individual measures depends on the time required for their development and implementation. Implementation of energy-saving solutions, e.g. the use of light-duty vehicles or change in user's behaviour and at least partial use of teleworking, is relatively easy and less cost-intensive projects. In the short-term perspective it can yield significant savings. The activities planned in the long term will require substantial investment in research and development of new technologies. However, taking into account the impact of e-mobility on the reduction of CO_2 emission continuing these activities seems justified from the point of view of sustainable development.

References

1. White Paper Roadmap to a Single European Transport Area – Towards a competitive and resource efficient transport system, COM (2011) 144, Brussels (2011)
2. EU transport in figures, Statistical Pocketbook, Luxemburg (2014)
3. Zachmann,G., Holtermann, M., Radeke, J., Tam, M., Huberty, M., Naumenko, D. and Ndoye Faye, A: The great transformation: decarbonising Europe's energy and transport systems, Bruegel Blueprint Series, Brussels, Belgium (2012)
4. Lapillonne, B., Pollier, P.: Energy Efficiency Trends in Transport in the EU. Enerdata (2015)
5. ODYSSEE MURE Project Partners http://odyssee.enerdata.net/nrd_web/site/
6. Motowidlak, U., Identification of transport fuel needs in terms of transportation, Materials Management and Logistics No. 4, pp. 19–28 (2014)
7. Filipowicz J., Kwiecień K.: Development of the electric propulsion of motor vehicles in the aspect of the environmental protection, Logistics No. 3 (2012)
8. Ślęzak M.: Polish ecological solutions motoring future?, http://superauto24.se.pl/ (12.05.2015)
9. Guziński J., Adamowicz M., Kamiński J.: Electric vehicle drive systems, Automatics-Electronics-Interferences No. 3 (2013)
10. Report on the technological and economic background of electromobility, its major stakeholders, research questions addressed, and expected project targets, www.project-emap.eu.
11. http://www.acea.be/press-releases/article/alternative-fuel-vehicle-registrations-28.8-in-in-first-quarter (15.05.2015)
12. www.cepik.gov.pl (01.03.2015)
13. AEA Energy Efficiency Trends in the Transport sector in the EU, AEA, ODYSSE MURE project (2012)
14. Schroten, A.: Behavioural Climate Change Mitigation Options, Domain Report Transport, CE Delft (2012)
15. AEA Energy Efficiency Policies in the Transport sector in the EU, AEA, ODYSSE MURE project (2012)
16. Tom Tom Traffic Index Reports, s. 64 <http://www.tomtom.com/lib/doc/pdf/2014-05-14%20 TomTomTrafficIndex2013annualAme-km.pdf (28.03. 2015)
17. Wappa P., Halicka K.: The importance and a source of transport congestion on the example of the Bialystok agglomeration, Economics and Management No. 4 (2011)

Pro-Innovative Systems of Decision-Making in TSL Enterprises: Behavioural Conditions

Joanna Fryca-Knop

Abstract TSL enterprises are constantly looking for development reserves, which would provide a relatively sustainable competitive advantage to support their development in an increasingly demanding environment. This advantage is possible due to the decision-making systems. The systems foster effective decisions that promote creation and implementation of innovations in the enterprise. Behavioural factors take important role when making such decisions. The recognition of their prominence is a must for an effective way to arrive at these decisions.

Keywords Decision-making system in the enterprise • Pro-innovative decision • Behavioural conditions

Introduction

The role and importance of innovation in today's activity of enterprises from the transportation–shipping–logistics (TSL) sector are strongly accented by theoreticians and practitioners as the key element allowing organisations not only to build sustainable competitive advantage, but also to survive in a complex and turbulent environment. As it is emphasised by the scholarship, their right to exist is not eternal, so they have to gain it over and over again [1]. For TSL enterprises operating in the twenty-first century, the innovation is the key to their development, and thus the condition of survival for these entities that operate "in the age of innovation" [2]. Thus, the key issues facing policymakers are the speed and effectiveness of decision-making. They contribute to the success of the managed entities. The objectives that are standing in front of them are the more difficult skills that require the ability to choose the ways of market expansion for enterprises operating in an increasingly complex and dynamic environment and which are subject to increasing pressure that aims to innovation. This means that the decision-making process is increasingly difficult.

J. Fryca-Knop (✉)
Faculty of Economics, University of Gdansk, Sopot, Poland
e-mail: j.fryca@ug.edu.pl

M. Bąk (ed.), *Transport Development Challenges in the Twenty-First Century*,
Springer Proceedings in Business and Economics,
DOI 10.1007/978-3-319-26848-4_16

Less and less decision-making problems can be programmed. They are often insufficiently structured, and decisions are made under conditions of conflict of interest and the lack of information. Therefore, in order to overcome the growing amount of difficulties rules for decision-making in the enterprise need to be recognized. These are, in particular, the mechanisms that govern the behaviour of decision-makers, and various social groups.

The main goal of this chapter is to determine the role of behavioural factors in making effective decisions that support the development and implementation of innovation in enterprises of TSL sector. The decision support systems exist in every business entity, so the very fact of their existence should not be attributed to the potential success factor. However, the practice of market activities of individual TSL enterprises indicates that they do not always operate in the way that ensures success. In order to identify the reasons for this state of affairs it is sought to look into the mechanisms governing the making of pro-innovative decisions.

Innovation as Changes Requiring Active Decision-Making in TSL Enterprises

The increasing complexity and turbulent environment necessitate more frequent changes of enterprises functioning in TSL sector in the course of doing their business. This is caused by the new market challenges, as well as the uncertainty and unpredictability of the future. The above observation applies not only to new enterprises, but also to the enterprises operating in the market for a long time. The regularity of changes establishes the need to constantly adapt the strategies either to the emerging opportunities or to the surrounding dangers.

Changes may be external or internal. The former are the result of the pressure coming from the business environment. They are most frequently expressed in the implementation of new products, the application of new technologies or modification of the existing organisational processes. The latter are inspired by the management of the enterprise. Although all of these changes are very important and necessary, those of innovative nature appear to be the most important. They guarantee technical progress and effectively reduce the risk of loss of competitive advantage. Thus, the search for and implementation of innovative change is an integral feature for the modern approach to do business effectively in the TSL sector [3].

At this point, however, one must ask whether all change means innovation. There is no doubt that if all innovation is a change, but according to Białoń not all change should be considered an innovation. She quotes Świtalski's features of changes that are indicative of their innovative nature of thinking that innovation is [4]:

- A result of modifying alteration or introducing entirely new elements to the way or the effect of the functioning of the enterprise
- A deliberate change, implemented consciously, of a lasting nature
- A change resulting in more efficient functioning of the entity or an increase in utility due to its functioning
- A change that should have the ability to spread out among other similar entities

It can be concluded that the concept of innovation is inextricably linked with the ones of change, news, reform or idea perceived as new. Moreover, it requires a corporate decision-making activity that takes into account the need for a systemic approach to the problem.

When deciding on innovation, one usually involves more than just one of the functional areas of the enterprise. Such situations, in the so-called pure form, are found only in the lower and middle levels of management [5]. At higher levels, decisions always have a mixed form because they must combine the contents of the problems in order to resolve various functional areas of such entities. They are, therefore, characterised by varying scope and result in different potential for innovation activities, the various resource capabilities, performance, innovative capabilities and organisational efficiency in this field [6].

They are made at different levels, which include the strategic, tactical and operational ones. The strategic decisions related to the choice of an appropriate strategy of innovation that will lead to decisions creating conditions for the development and implementation of innovations are particularly important in the practice of making decisions in TSL enterprises concerning innovation. These decisions are crucial for the success of enterprise in the market and determine prospects for development.

They are specific and should be something other than a choice that constitutes the basic effect of the decision-making process. Making decisions endowed with the strategic dimension requires involvement of a decision maker, for whom the future is the most important. For this reason, all actions taken should take this priority into account [7]. Due to the long time perspective of the choices made, these decisions are taken at the highest level of uncertainty and risk. Furthermore, they offer the highest level of significance effects in both economic and social aspects. They induce the fear of change, and even active resistance to their introduction because their implementation often requires actions such as a violation of the existing relations of power and influence, a different allocation of limited resources as well as changes in organisational culture [5]. For this reason it is important to recognise the rules by which they are made so that the TSL enterprises were able to guard against errors and improve decision-making systems and meet the increasing demands of management.

Place of a Decision-Making System in the Enterprise

The systemic thinking on the functioning of enterprises requires the ordering of decision-making and comprehensive treatment of issues related to them. Hence, in order to make an enterprise function effectively, one must give due attention to the conditions and tools allowing for making optimal decisions. Yet, since the economic optimisation is a conditional process, these entities should adjust their sphere of regulation and decision-making systems to their specific nature [8]. The information and decision-making systems are deliberately organised structures of people, equipment and procedures for collecting, processing and distributing information. These are done for the purpose of decision-making centres in the enterprise [21]. They should have universal functional structures and, at the same time, coordinate activity of functional regulators in the enterprise at all functional levels. The system of decision-making is both feasible and regulatory. The feasibility of the system is reduced to taking into account behavioural factors associated with the person or the decision maker and various groups involved in the decision-making processes of non-quantifiable nature. They all determine its efficient functioning. This system is a set of factors that determine the final shape of managerial decision.

In this system the transformation of economic events is made. Both typical and incidental impulses must be treated as economic interventions and the optimal managerial decisions. In order to make choices most favourable in these circumstances, the system should use many decision concepts that are worked out by specialised substance regulators. In the operating conditions of the enterprise, from the perspective of decision-making system, its system of supporting the choices is multidimensional and strongly selected under the influence of selected types of events. This will be subject to the transformation of decision. One has to take into consideration that this nature of optimal decision is given only in the system of decision-making. Thus, the basic problem is the selection of the structure of the elements of the system. The selection and ranking of elements that are relative to each other shape the level of efficiency of the decision-making process and the nature of the optimal decision. This affects, therefore, the efficiency of the entire enterprise. Constructing a specific enterprise in decision-making system encounters a complication. It results from a variety of regulatory and diversity of social conditions which may occur as a result of changing economic realities. But the system ability to generate such decisions ultimately verifies the design solutions. This contributes to a positive verification of the organisational behaviour of a market enterprise.

When compared with each other, the decision-making systems show significant differences in the level of complexity and sensitivity to changing environments. This is particularly true when practical circumstances of individual enterprises are taken into consideration. One can note, however, significant similarities in the selection of decision factors in the so-called subject section. Their configuration includes all of the events and circumstances that will occur in these economic entities. Each element of the system works according to the logical, analytical and normative schemes. Incidental event factor treats each economic event requiring a decision fully individually in terms of the capabilities to perform certain actions. At owned by the enterprise production capacity and the legal and financial conditions, performance factor examines the level of cost-effectiveness of these measures. The sociological factor takes into account the impact of the sociological and social ties and the psychological factor shows the impact of the psychological characteristics of the decision maker on the final shape of managerial decision [8]. At the same time each of them follows to its own logic. The processes taking place in these elements have the only proper conduct according to the well-known and recognised limitations of decision rationality.

To make the best decision in the enterprise, basic requirement of the optimisation must be met. As such one may consider the variant of a selection that must be positively confirmed by the decisive factor called an incidental information factor. Search for a solution of the decision problem is made at the following levels: internal, related to the production potential, organisational and human of the enterprise and external related to the environment conditions [10]. For this reason, it must confine in the so-called potential field of decision. To some extent it co-creates the qualitative factors. When one of the variables that describe partially the potential field of the decision prevents accomplishment of the assigned economic task that is determined by a decision variant, the enterprise has either to involve the negative managerial decision or to take into account the analysis of another variant of choice. The fulfilment of conditions of this field may be enough to make the final decision, but usually a decision variant is subject to valuation by other optimisation factors [8].

The second field of decision discretion that could be called an acceptable field of decision creates a set of decisions ensuring the realisation of the economic characteristics considered as desirable or model ones. The detailed examination of economic feasibility of implementation of actions which should be taken in the case of admission of realisation of the following tasks occurs in a productivity factor. It should absolutely deny the decision variant. This results in inefficiency that may occur in any form. The final decision, however, can be deformed by other factors of the decision-making system.

Transformation of decision variants in the final managerial decisions in the first part is subordinated to the rigours of selection fields. They are created by deterministic factors that are the incidental information factor and productivity factor. Yet, the concepts of systemic decision-making developed at these stages are liable to further changes. These results appear in any enterprise with probable social conditions that are commonly identified as sociological and psychological factors. Their significant impact on decision-making in the enterprise has made the behavioural approach to the organisation research. Specific behaviour of individuals and groups occupies a very important place, and research directions of their interactions are conducted today. These factors are referred to as behavioural.

It is particularly the social factor which assumes that the enterprise is also a social creation. It is, along with the effectiveness criteria, conditioned by the specific function within society. One cannot create knowledge-based organisation without a common vision. When focused on the target, the enterprise requires overcoming resistance from inside the organisation. This results in reluctance to new solutions and to exploit the potential of people. The loftiness of purposes stimulates creative thinking and finding new ways of doing things. Hence, there is the common vision of breaking the resistance at the planning stage, which is the structure of decision variants and not in the implementation phase [11]. This fact has obvious benefits.

Any innovation-oriented decision-making in an increasingly changing and complex economic environment turns out to be an increasingly difficult process. Enterprise set for success must conform to international standards of efficiency, and reactive actions are not enough: "actions that are ahead of their time are needed, generating a new vision of the existing, inspiring the creation of new markets, including at least the seeds of innovative solutions that are difficult or impossible to be imitated by the competitors" [12]. It can be concluded that each enterprise should be pro-innovative, and for this reason it is necessary to involve the flexibility and courage in taking the risky and often unconventional market actions. It should be emphasised that the decisions are affected not only by the objective conditions related to the nature of the problems and the impact of external factors, which include the incidental information factors and efficiency, but also the social context (sociological factor) and the characteristics of the decision maker (psychological factor).

The Role of Behavioural Factors in the Decision-Making Pro-Innovativeness in TSL Enterprises

The social context of pro-innovative decision-making process can refer both to the climate that accompanies the decision-making and implementation of innovation and their collective elaboration. In the latter case, it is worth noting the synergies resulting from the strengthening of individual potential of decision makers. The behaviour of individuals in the group is undoubtedly much more than just the sum of individual actions—it is some kind of reinforcement of individual actions of people involved in the decision-making processes [13]. This approach shows a situation in which the person making a decision under partial function becomes a participant of the decision-making formal group (functional or task) or informal.

Each of these groups has their specific objectives that are consistent with the objectives of enterprise or in conflict with them. For this reason the partial decision of a sociological factor should be determined by the purpose of groups in the enterprise and the target of the enterprise. Thus, the decision process in enterprises within sociological factor can take four forms. These are the following:

- When a decision is worked out by the managerial group, it means that the choice is made by a particular person, but as a result of a direct equivalent using of the mental concept of co-creating it persons
- When a decision is made with the help of expert groups involved in the decision-making process as a source of information and opinion on the decision situation and the consequences of those choices
- When the decision is made with the participation of all employees participating in the whole decision-making process or in some stages of it
- When a decision is made individually or as a group, however, it remains under the actual or potential impact of people, who in a specific way affect the decision maker or decision-making group, as the so-called pressure or support groups

As the implementation of decisions on innovation depends to a significant extent on the level of their acceptance by the organisational or non-organisational community, decision-making system should be set up in such a way that even the most daring market activities should have a chance to happen and to be implemented into real life. Usually such decisions are accompanied by high risk and uncertainty.

Therefore, it is important to create an organisational climate conducive to innovation. The main function to stimulate innovation readiness performs organisational culture as a source of "creative climate" for innovation, shaping the attitude of employees towards them. As indicated by empirical research, it should be focused on [9]:

- Creation of environment conducive to change
- Development of knowledge and skills
- Development of tolerance for risk and uncertainty
- Strengthening of respect for the non-conformist attitudes
- Fostering the climate of respect for innovators

This kind of social context will contribute to the elaboration by the TSL enterprise formulas of innovation-friendly choices. Furthermore, the group formation of decision creates conditions for collaborative learning and thus for creating the knowledge in the enterprise. Collaborative creation of knowledge has two dimensions in the enterprise. Firstly, it is necessary to rethink analysed problems, which requires the ability to connect potential of many minds, so as to establish a common potential greater than the potential of a single mind. Secondly, a coordinated action is required in the frame of implemented innovations, which builds operation trust in members of the team. This means that each team member is aware of the actions of others, and this again allows creating the feeling of safety to avoid an error [11]. The role of sociological factor in the decision-making process is activated in the following forms:

- Social attitude to the situation of decision-making
- Social attitude to the environment
- Social attitude to risk
- Perceived by a group of its role in decision-making process
- Force of social impact

However, not all decision-making processes take place in a team. Some of them eventually are verified by a single decision maker or even some are taken individually. The psychological factor is the second among the behavioural factors that are functioning in the decision-making system in the enterprises that are typically the last link in the creation of the concept of making choices. It is therefore worth considering what characteristics can be described as a decision maker. Then, it should be attempted to answer the question what the quality of their design allows for more effective pro-innovating decision-making. Behaviour of the decision maker in the decision-making processes are the result of his or her:

- Attitude to the situation of decision-making
- Attitude to the environment
- Attitude to risk
- Perception of the nature and direction of pressure on him or her

The decision maker, as a system that chooses, has a set of relatively stable characteristics, which are the source of his or her behaviour. To be recognised, it is worth to use the sequence of the innovation process proposed by Bal-Woźniak [14], such as awareness, readiness, skills and abilities. It is a modification of the well-known formula of innovative K-M-W (know-may-want) submitted by the author as a K-W-A-C (know-want-able-can) [15].

Factors of the first group—the "Know" type—are associated with consciousness accompanying the decision maker innovation processes. It is a state of mind, so that he or she is aware of the internal phenomena of the organisation and phenomena occurring in the external environment, while being able to react to them. One can say that this part refers to the ability of recognising the situation requiring action [14].

Awareness of the innovation needs is therefore the decisive element of perceiving the innovative nature of the situation as signalled by the stimuli. At this point it is worth noting that innovation can be attributed to different places where they start, ideas that were the inspiration as well as the causes of this phenomenon [6]. In each case, the source of innovation requires the relevant threshold of consciousness which is at the basis of knowledge of the need, necessity or chance to take initiatives aimed at creation and implementation of innovation. It is the base threshold of decision maker pro-innovation activity. It can manifest itself in both his or her attitude to the situation of decision-making as well as sensitivity to the direction and strength of the social impact on the decision maker. Thus the level of preparedness to take pro-innovation decision may be determined.

Willingness to take action is a determinant factor groups like the "Wanted" type. It conditions innovation arising from the consciousness of the validity of a specific problem (as a result of the impact of the first group of factors). It means decision on something, desire and inclination [14].

There is no doubt that the willingness to engage participants of an organisation, both individualised and group decision maker, depends on their level of motivation to undertake activities in the field of exploration and realisation of innovation in the enterprise. Basically, the two groups should be mentioned as reasons that ultimately shape his or her behaviour: self-motivation and external motivation [16]. The former may be characterised by two issues derived from the structure of personality:

- Expectations of the individual in relation to the usefulness of the result achieved through the implementation of the action taken
- Unit conviction as to the feasibility of the measures taken, or the assessment of the likelihood of achieving the intended result

In the case of external motivation one should pay attention to the action of team incentives. They not only apply to shape the attitude of the decision maker, but also are a substantial area of innovation-oriented decisions. They allow the targeting behaviour of employees and shaping their attitudes towards innovation. These stimuli include all interactions that give certain social objectives gratifying value. Stimulus construction itself allowed shaping the innovation attitudes of the employees and their teams result mainly from the rules adopted by the incentive system of a specific enterprise. Pro-innovative motivating requires the use of both economic and non-economic incentive instruments. Among them the following are the most important [17]:

- Systemic approach to motivate innovative employees
- Motivating aimed at building commitment
- Shaping the level of remuneration at a relatively high level, competitive on a regional basis
- Rewarding effects
- Using in a wide range of different immaterial instruments tailored to the expectations of benefits and combining them with the material instruments
- Flexibility in remuneration schemes

Yet, an analysis of the literature on the factors that motivate employees to innovative behaviour goes beyond strictly related to motivation and tends to emphasise the role of organisational commitment. Empirical studies [20] indicate the relationship between innovation of the employees, particularly the decision maker, and their strong identification with the organisation, expressing the affective commitment, and a negative relationship with calculative commitment. The development of the sense of belonging and the loyalty of employees are thus important elements that stimulate their willingness to engage in innovative processes.

A very important factor is the willingness to take risks, which is undoubtedly related to the innovation-oriented decision-making. Decision makers have various approaches to risk. Hence, it can be noted that the differences in their behaviour concerning the possibility of implementing innovations often inhibit the activity of TSL enterprises in this field. This was also confirmed in the study conducted in logistics companies in the Polish region of Podlasie, where the fear of risk, apart from the lack of support from management, was indicated by respondents as one of the main factors limiting the implementation of innovative solutions [18].

The most important meaning in the course of the innovation process is attributed to factors like "To be able to" associated with some fluency of conducting innovative business. Ability to react quickly to changes in the environment through the creation and implementation of innovations in the various functional areas of the enterprise has become a challenge that is not easy to cope with. It is essential to learn continuously, as an organisation, creation and using of knowledge, treated as one of the business functions. The most desirable features of modern TSL enterprise aspiring to become the innovative one are the decision makers' competences, particularly their development abilities, learning ability, creativity and the ability of forecasting, response and manufacturing [19].

Specific abilities, which decision maker should have, depend on the situation. For example, the ability to respond in the direction of change between the market and management (product innovation) or between the organisation and operation (organisational innovation) requires connection between the development abilities in the field of market development and management strategies. Besides, the ability to learn in terms of organisational and operational changes is essential in this case. By combining ability to learn in terms of organisational and operational changes of creativity in terms of knowledge and systematic thinking, the decision makers have the ability to create knowledge and innovative products and so on. However, all the above abilities have to be rooted in the respective competence of decision makers, especially in their creativity.

The last element that is worth noting consists of the group of factors like "Be able". They refer to the possibility of realisation of innovation in the enterprise. The effectiveness of both individual (psychological factor) and collective decision maker (sociological factor), and the social context of innovation (sociological factor), is conditioned by an organisation. One can extract multiple conditions necessary for innovation to appear. Among them one may find financial, technical, organisational, systemic conditions and other such [14]. It should be emphasised that the impact of behavioural factors in decision-making systems of TSL enterprises is also the function of the ability (deterministic factors) defining the base threshold of optimisation for pro-innovation decisions.

Decisions on creating and implementing innovations are rarely simple. They require constant intellectual effort and reconcile the interests of different social groups.

It is not an easy task. That is why the enterprises—in which situations of taking wise decisions are repeating—are of a great interest to researchers. The effectiveness of influences of behavioural factors does not result solely from the ability of correct understanding of decision makers and their intelligence. In addition to this, there should be deep organisational wisdom, especially the one determined by the wisdom of decision makers. A set of their features and capabilities lead to better decision-making.

Conclusions

In conclusion it can be said that the modern economy creates exceptional conditions for the functioning of TSL enterprises. Their needs must be reflected in the well-functioning decision-making systems. These systems should primarily enable the enterprises of this sector to take pro-innovative decisions focusing on their market success, allowing them for development in the long term. However, for the development it is not enough to have more and better information and the availability of latest decision-supporting systems, but also the ability to use them by the enterprise is required, as well as an organisational reflexivity, self-awareness and a favourable social climate. It should be emphasised, therefore, that the form of the decisions made in the TSL enterprises affects not only the deterministic factors influencing potential and acceptable field of decision, but also the probabilistic conditions associated with the impact of behavioural factors. Their role in making effective, pro-innovative decision is expressed in creation of deep organisational wisdom of complex and changeable market processes, acceptable methods of organisational behaviour, strengthening creativity and innovation, increasing the speed of response to market opportunities and threats and minimising the resistance against changes due to taken pro-innovative decisions.

References

1. Simons, R.: Levers of control: how managers use innovative control systems to drive strategic renewal, pp. 114. Harvard Business School Press, Boston (1995)
2. Jansen, F.: The age of innovation, pp. 3. Prentice Hall, London (2000)
3. Wysocki, J.: Innowacje jako warunek rozwoju współczesnych przedsiębiorstw, In P. Niedzielski, K. Poznańska, K.B. Matusiak (Ed.): Kapitał ludzki - innowacje - przedsiębiorczość. Zeszyty Naukowe Uniwersytetu Szczecińskiego, No. 525, pp. 23–34. Wydawnictwo Naukowe Uniwersytetu Szczecińskiego, Szczecin (2009)
4. Białoń, L. (Ed.): Zarządzanie działalnością innowacyjną. Placet, Warszawa (2010)
5. Bolesta-Kukułka, K.: Decyzje menedżerskie. PWE, Warszawa (2003)
6. Janasz, W., Kozioł-Nadolna, K.: Innowacje w organizacji. PWE, Warszawa (2011)
7. Penc, J.: Decyzje menedżerskie – o sztuce zarządzania. Wydawnictwo C.H. Beck, Warszawa (2001)
8. Szałucki, K.: Przedsiębiorstwa transportowe. Warunki i mechanizmy równowagi. Wydawnictwo Uniwersytetu Gdańskiego, Gdańsk (1999)

9. Knap-Stefaniuk, A.: Innowacje a konkurencyjność przedsiębiorstw. Zarządzanie zmianami, No. 6 (2007)
10. Letkiewicz, A.: Samoregulacja w symetryzacji systemów przedsiębiorstw. Wydawnictwo Uniwersytetu Gdańskiego, Gdańsk (2013)
11. Senge, P.M.: Piąta dyscyplina. Oficyna Ekonomiczna, Kraków (2004)
12. Perechuda, K. (Ed.): Zarządzanie wiedzą w przedsiębiorstwie. WN PWN, Warszawa (2005)
13. Majecka, B.: Zachowania organizacyjne przedsiębiorstw, pp.56. Wydawnictwo Uniwersytetu Gdańskiego, Gdańsk (2002)
14. Bal-Woźniak, T.: Innowacyjność w ujęciu podmiotowym. Uwarunkowania instytucjonalne. PWE, Warszawa (2012)
15. Zajączkowski, M.: Podstawy innowacji i ochrony własności intelektualnej. Economicus, Szczecin (2003)
16. Reykowski, J.: Teoria motywacji a zarządzanie. PWE, Warszawa (1975)
17. Sajkiewicz, A., Woźniakowski, A.: Proinnowacyjne wynagradzanie – badania IPiSS. Zarządzanie zasobami ludzkimi, No. 2 (2009)
18. Dziekoński, K., Chwiećko, J.: Innowacyjność przedsiębiorstw branży TSL. Economics and Management, No. 2 (2013)
19. Baruk, J.: Zarządzanie wiedzą i innowacjami. Wydawnictwo Adam Marszałek, Toruń (2006)
20. Jafri, M.H.: Organizational commitment and employee's innovative behaviour. Journal of Management Research, Vol. 10, No. 1 (2010)
21. Penc, J.: Leksykon biznesu. Agencja Wydawnicza "Placet", Warszawa (1997)

Simulation of Technical and Economical Processes as an Initial Phase of Electric Bus Fleet Implementation to Operation in Urban Public Transport Company

Krzysztof Krawiec

Abstract Analysing the possibility of electric, battery-powered bus operation one should take into consideration many factors, e.g. the plan and profile of the bus routes, and technical, operational and economical characteristics of the public transport company. This chapter presents the concept of using .xml format files as a form of writing technical, economic and operational characteristics of battery-powered buses. The purpose of their use is parameterization of the universal model which analyses the possibility of introducing electric buses into service for potential user, e.g. the public transport company or transport organiser. This chapter presents the characteristics of the .xml files relating to the public transport network and timetable, operational and technical characteristics of the enterprise, infrastructure, transport business economics and issues related to the environmental aspects of the use of buses equipped with different types of drives in public transport.

Keywords Electric bus • Urban public transport • Electromobility

Introduction

Currently, the vast majority of the public transport fleet consists of buses with diesel engines. For a few years substantial efforts have been made to replace the fleet of buses with diesel engines with buses showing better indicators from the point of view of environmental protection. Such actions should be associated with increasing European Union restrictions concerning emissions of CO_2 and other dangerous substances. The position of the European Commission on the electric vehicles was presented in 2010 in the document "A European strategy on clean and energy efficient vehicles" which encompasses the strategy on the promotion of ecologically

---

K. Krawiec (✉)
Silesian University of Technology, Gliwice, Poland
e-mail: krzysztof.krawiec@polsl.pl

M. Bąk (ed.), *Transport Development Challenges in the Twenty-First Century*,
Springer Proceedings in Business and Economics,
DOI 10.1007/978-3-319-26848-4_17

clean and energy-efficient vehicles (cars and buses) [1]. The strategy concerns both the vehicles equipped with conventional diesel propulsion and those using new propulsion technologies (hybrid, electric and hydrogen propulsion) which will be introduced. Furthermore, in particular for electric-powered vehicles certification rules and technical standards will be implemented, in order to make it possible to charge these vehicles and connect them to the electric network in the whole EU using the typical charger.

Aspects of Electric Bus Implementation

The technology of e-buses already exists; however it requires further development in order to meet the needs of users—public transport companies and passengers. While analysing the possibility of implementation of electric, battery-powered buses to urban public transport many factors should be taken into account [3]:

- Operational aspects of implementation
- Technical problems that impede the implementation of electric buses
- Influence on the environment
- Economic costs of different types of propulsion

In the following part of this chapter these aspects are presented and explained in the context of their further application.

Operational and Technical Aspects

Implementing battery-powered electric buses is a challenge from the operational point of view. Because of the technical limitations of e-buses two variants are possible:

- Adjustment of the schedule to new electric fleet
- An attempt to conform effectual bus schedule

Any modification of the timetable should be minimised due to the fact that passengers are accustomed to effectual bus schedule. When possible one should try to conform effectual bus schedule and this variant will be analysed in this chapter. Current timetable (schedule) should thus be modelled as a constant keeping the option of changing the timetable open. E-buses are characterised by technical limitations such as:

- Limited range caused by imperfect battery technology and the consequent need for recharging the battery during the route
- Necessity to purchase and install additional technical infrastructure such as battery charger (plug-in, pantograph or inductive) or battery exchange points

Currently used energy source are lithium-ion batteries or lithium–iron–phosphate with high power. The main factor which limits their use, from the technical and operational point of view, is the limited range of buses. Another limitation is the variation of their performance depending on different weather conditions; that is, the range of the battery-powered bus depends on temperature. As a result, there appears the problem of necessity to recharge or replace the battery on the route to provide the operability of the bus throughout the day. In order to enable modelling variation of weather conditions one can use historical data for the corresponding periods which can be obtained from meteorological stations. These limitations are important from the point of view of the necessity of providing full operability for electric buses which is needed by the final customer (public transport company).

Network of public transport bus company is strictly connected with operational and technical aspects of analysing implementing of electric buses. E-buses, in contrast to buses equipped with diesel combustion, are much more liable to variations of route profile on which they operate. Aspects such as latitude, longitude and elevation (above sea level) should be taken into consideration while modelling the possibility of replacing the diesel buses with e-buses. Each section between two bus stops is characterised by different parameters: plan and profile, and energy consumption for subsequent types and models of buses (quantity of fuel for conventional buses; kWh for electric buses). The quantity, type and distribution of technical infrastructure are the subject of optimisation. Surplus technical equipment generates extra costs for public transport company. The scarcity of the technical equipment, however, causes operational problems.

Ecological Aspects

The influence of different types of propulsion on environment is the major determinant of the choice of particular type of buses. Emissions of the following substances, carbon dioxide (CO_2), sulphur dioxide (SO_2), nitrogen oxide (NOx) and suspended particulate matter (SPM), and their influence on human health must be taken into consideration. Each type of propulsion (diesel, hybrid, gas, electric propulsion) is characterised by different level of emissions. Due to the European Union policy in terms of environment protection such as successive Euro norms for combustion engines and offset policy (emissions trading) the amount of pollutants being emitted by buses during the operation is a subject of interest of public transport companies.

Carbon dioxide and other harmful substances are emitted not only during operation of the bus but also in the process of electricity production. The aim should be to increase the share of renewable energy sources and minimise emissions at the level of electricity production. The comparison of global CO_2 emissions in Polish and German energy mix is presented in Table 1.

Even in ecologically cleaner (from CO_2 emissions point of view) German energy mix total emissions of carbon dioxide are still higher for electric buses than for diesel buses. Further increase of the share of renewable energy sources in energy mix would however decrease total CO_2 emissions. The variant of 100 % renewables should also be taken into consideration.

Table 1 The comparison of global CO_2 emissions for electric buses and conventional diesel buses

	Electric bus (length = 12 m)	Diesel bus (length = 12 m)
Energy consumption	192 kWh/100 km	38 l/100 km
CO_2 emission in German energy mix	110.6 kg CO_2/100 km per kWh	99.9 kg CO_2/100 km per kWh
CO_2 emission in Polish energy mix	172.8 kg CO_2/100 km per kWh	99.9 kg CO_2/100 km per kWh

Source: own study on the basis of Evo Bus company materials

Economic Aspects

One of the fundamental aspects of deciding on the choice of the bus fleet to handle the communication lines within the public transportation is the economic aspect [5]. This also applies to the bus fleet with electric drive. The main components of the costs associated with the implementation of electric-powered buses for public transport are as follows:

- Costs of acquisition
- Operating costs
- Costs of technical infrastructure
- External costs

Electric buses are characterised by a higher cost of acquisition than the conventional ones and additional technical infrastructure (recharging or replacing battery points, flywheel) which is not necessary in traditional combustion propulsion [2]. E-buses have relatively low operating costs, arising from the characteristics of the electric drive.

Simulation

The author of this chapter is a member of the "Models and Methods for the Evaluation and the Optimal Application of Battery Charging and Switching Technologies for Electric Busses (CACTUS)" research project which is being conducted at Silesian University of Technology.

Sub-models of the Simulation

The simulation was divided into the following sub-models [4]:

- Transportation model concerning bus route network, bus stops, schedule, vehicle operation plan

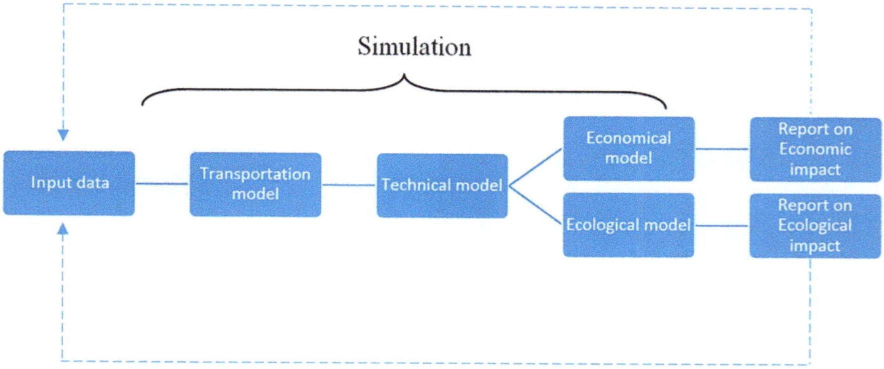

Fig. 1 Scheme presenting successive stages of the model (source: own study on the basis of CACTUS materials)

- Technical model concerning drive, technical and behavioural characteristics and energy consumption of the fleet (e.g. charging strategy) as well as inclinations of network sections (nodes, sections)
- Ecological model concerning emissions of harmful substances (CO_2, SO_2, NOx, SPM) depending on the type of propulsion
- Economic model concerning subsequent types of costs

Successive stages of the simulation are presented in Fig. 1.

In the simulation input data are being processed in successive sub-models: transportation and technical. In further part, economic and ecological sub-models are being processed simultaneously. Reports regarding economic and ecological impact are the result of the simulation. In case of unsatisfactory results, the feedback goes back to input data stage.

Initial Phase of the Simulation

Input data should be stored in a format providing clarity and the possibility of dynamic swapping the data in case of feedback in the end of simulation. The format that meets these requirements is Extensible Markup Language (XML) which is both human readable and machine readable.

Scheme shown in Fig. 2 presents the assignment of input data to the respective sub-models. Three XML files (Network.xml, Timetable.xml and Operation.xml) represent transportation model, Infrastructure.xml file represents technical model and Economy.xml and Ecology.xml represent economic and ecological model, respectively. These XML files are batch files for simulation algorithm. The structure of the XML files will be explained in the example of Economy.xml data stage if the result of the simulation is unsatisfactory.

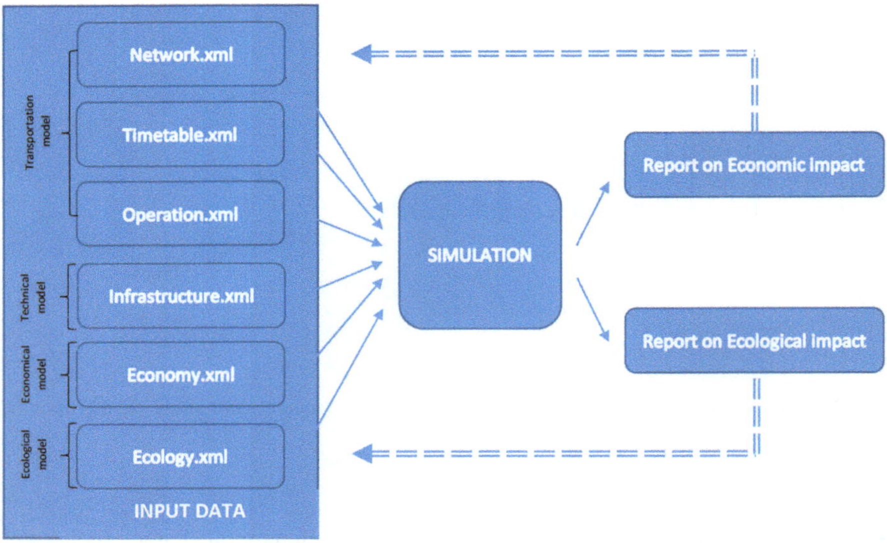

Fig. 2 The assignment of input data to the respective sub-models (source: own study on the basis of CACTUS materials)

Economy.xml as an Example of XML Structure

The file contains definitions of different types of costs: acquisition costs, operating costs, infrastructure costs and external costs which are enclosed in the tag <CostType>. The structure will be presented in the example of acquisition costs. Operating costs, infrastructure costs and external costs have been written in similar way:

```
<CostType id="1" name ="Acquisition Costs" variable="KA"> <!--
present value of the acquisition costs of bus -->
     <Category name="General Costs">
     <Property key="Double-layer Capacitor Cost" unit="EUR"
variable="K_kond">0</Property>
  <Property key="Subsides For Bus" unit="EUR" variable="DA">1.302</
Property>
  <Property key="External Finance Rate" variable="u_kred_A">0.5</
Property>
     </Category>
     <Category name="Annuity Costs">
  <Property key="Credit Period" unit="year" variable="N">5</
Property>
  <Property key="Loan Interest Rate" variable="s">0.07</Property>
  <Property key="Market Interest Rate" variable="i">0.02</Property>
```

```
<Property key="Repayment Term" unit="year" variable="M">5</Property>
     </Category>
   </CostType>
  [Example of Economy.xml structure. Own study.]
```

General acquisition costs are the costs of the bus fleet: the cost of the bus itself and the cost of double-layer capacitor taking into account subsides and external finance rate. The units for costs are Euro (EUR), which will be later converted into Polish currency (PLN). The category “annuity costs” refers to financial conditions of the investment in electric buses. Credit period, loan interest rate, market interest rate and repayment terms are taken into account.

Necessary economic values for three types of energy storage are given in the “EnergyStorageType” tag. Costs and lifetime of the battery (and spare battery) as well as the capacity of the battery are important from the view of further calculations in the model:

```
  <EnergyStorageType id="1" name="Battery">
      <Category name="Extended Properties">
  <Property key="Capacity" unit="kWh">0</Property>
  <Property key="Cost" unit="EUR">0</Property>
  <Property key="Lifetime" unit="year">10</Property>
<Property key="Spare Battery Cost" unit="EUR">0.306</Property>
      </Category>
  </EnergyStorageType>
  [Example of Economy.xml structure. Own study.]
```

More properties of energy storage types which include battery, ultracapacitor and flywheel are presented in Operation.xml file. The file contains such properties as manufacturer and model of the bus, possibility of charging (including charging power, time and efficiency) and exchanging energy storage types. The above-mentioned category “Extended Properties” of Economy.xml refers to these important from economical sub-model properties of the energy storage types which were not included in Operation.xml files.

Summary

Implementing buses with alternative propulsion to urban public transport is a long-lasting process which can be supported by the simulation. Consecutive sub-models concern different aspects of the implementation: operational, technical, ecological and economic. Presented fragments of input files written in Extensible Markup Language (xml) provide dynamic edition by human and are machine readable. Thanks to the economic sub-model, forecasting and promoting of bus fleet exchange to electric buses which do not emit CO_2 in the area of exploitation are more effective.

References

1. A European strategy on clean and energy efficient vehicles (2010), Brussels, 28.4.2010 COM(2010)186 final.
2. Bąkowski, W.: Procesowe podejście do ustalenia granicy opłacalności eksploatacji taboru w transporcie lądowym, Przegląd Komunikacyjny no 5 (2006).
3. ERA-NET Electromobility+: „Models and Methods for the Evaluation and the Optimal Application of Battery Charging and Switching Technologies for Electric Buses (CACTUS).
4. Markusik, S, Krawiec, S. , Łazarz, B., Karoń, G. , Janecki, G., Sierpiński, G.: The technical and operational aspects of the introduction of electric – powered buses to the public transportation system, Transport Research Arena, Paris (2014).
5. Wyszomirski, O.: Transport miejski. Ekonomika i organizacja, Gdańsk (2008).

The Impact of Organizational Culture on Bicycle Commuting Frequency: The Research Based on Example of Three IT companies

Romanika Okraszewska

Abstract This chapter describes a study on work-related factors affecting employees' decision on bicycle commuting. Employees' transport choices are related to organizational culture. This chapter refers to E. Schein's definition of organizational culture model. The study was made in order to determine the impact of specific levels of organizational culture on the transport behavior of employees. It is based on a survey performed in three IT companies. The main objective in the selection of companies for the research survey was to eliminate differences in other relevant factors affecting the employees' choices. The results of the surveys, supported with personal interviews, indicated that in the case of simultaneously displaying opposing behavior at the first and second levels of Schein's organizational culture model, the values have a stronger impact on the employees' transport choices than artifacts. The results can be applied to increase the effectiveness of mobility plans if they are being implemented in companies.

Keywords Bicycle commuting • Organizational culture • Mobility management

Introduction

In the light of environmental threats caused by motorized means of transport, new challenges have emerged: to manage social transport choices, to change current transport patterns, and to create a new mobility culture [1]. To meet those objectives, walking and cycling are the most likely candidates. Meanwhile, the modern world is car dependent.

A lot of research has been done in order to understand commuting behavior and the reasons for choosing one means of transport over another. This chapter

---

R. Okraszewska (✉)
Gdansk University of Technology, Gdansk, Poland
e-mail: romanika.okraszewska@pg.gda.pl

M. Bąk (ed.), *Transport Development Challenges in the Twenty-First Century*,
Springer Proceedings in Business and Economics,
DOI 10.1007/978-3-319-26848-4_18

describes the study originating in an analysis of the factors which are related to organizational culture and possibly affect the choice of bicycle commuting. The study is based on the assumption that not only artifacts but also values and assumptions constituting the company's organizational culture affect the employees' transport behavior.

The following text is structured as follows: The next section presents the state of the art of work-related factors affecting bicycle commuting. The third part introduces the motivation for the research undertaken and formulates hypotheses on the effects of specified levels of organizational culture on travel mode choices. The fourth part explains sample determination which is based on unifications of some important factors described in the second part while the fifth part is focused on the method of data collection. The sixth part presents and discusses survey results. The seventh part puts forward the hypothesis validation, summarizes key findings, and outlines possible directions for future research.

Common Work-Related Factors Affecting Commuting by Bicycle

The bicycle can be used for recreation or as a means of transport. The factors which encourage people to use bicycles for concrete purposes are subject to research. The literature shows that the elements that influence bicycle use as a practical travel mode can be grouped into users' characteristics, trip characteristics, environmental context, facilities, and subjective perceptions [2, 3]. Trip destination is a relevant factor when understanding bicycle use. Travels related to commuting are the main sources of traffic in the cities [4]. Therefore, changing transport patterns of employees would have a great significance for urban traffic. The decision to commute by bicycle might be viewed as a three-tier process: initial considerations–trip barriers–destination barriers [5]. The typical barriers occurring at the destination are not insurmountable, but could be crucial [5]. Therefore the conclusive impact on the decision to commute by bicycle or other means of transport could be from the work-related factors. However, although researchers have shown great interest in cycling in general and in bicycle commuting especially, little attention has been given to the effect of work-related factors on transport choices. Heinen et al. [6] describe the relation between levels of bicycle commuting and the following factors: sector, job position, terms of employment, number of hours at work, clothing, displacements needed during work, transport of large or heavy objects, number of working days, number of working hours, number of working locations, and colleagues' opinions. In addition to this, Wardman et al. [7] mention location of the workplace, availability of public transport, and facilities at the workplace.

The fact that availability of showers or clothing lockers at the workplace does not appear to inspire bicycle users to commute by bicycle more frequently is important for further considerations [8]. Gender correlation is repeated in many studies; men are more likely to bicycle to work than women [8]. The opinion of colleagues is described as a statistical significant variable in transport decision model by

Heinen et al. [6]. Explicit consideration of psychological factors might help us understand the process of making individual decisions concerning the means of transport [9]. Psychological factors including perceptions, identity, social norms, and habits are used increasingly to understand travel mode choices [10, 11]. In this context, one subjective factor should be taken into account—the effect of peer group travel habits on individual choices [5]. The co-workers are a natural reference group for those employees, whose values and standards of behavior are adopted as a model for their own behavior [12].

Background and Hypothesis Formulation

The observation that the proportion of workers who commute by bicycle to work in three different IT companies is fundamentally different and depends on the company they work for was the main inspiration for this study. The research was additionally motivated by the fact that the corporations do not significantly differ in aspects that are commonly identified in scholarship as crucial for the transport decision making. This study is based on the assumption that a company organizational culture influences the transport behavior of employees. For the forthcoming argument the definition of E. Schein's organizational culture model was chosen as the most appropriate. In this model of organizational culture three levels are highlighted: artifacts, values and norms, and basic assumptions [13]. An external observer has a different perception of particular levels. Artifacts are visible to the employees but also visible and recognizable for external parties. Basic assumptions are deeply embedded in the organizational culture. They are experienced as self-evident and unconscious behavior and hard to recognize even from within. With the help of scholarship on factors affecting commuting by bicycle, the psychological determinants of transport choices and private observations, the hypothesis was formulated as follows: **There are invisible elements of organizational culture which have a crucial influence on the employees' transport choices**.

Sample Determination

The choice of employees from the IT companies was a natural consequence of the personal observations on the differences in travel behavior in this group. The second selection criterion of companies was the will to unify, as far as possible, non-work-related factors affecting bicycle use for commuting. In addition to the sector, the company selection criterion included age structure, gender structure, size of branch, and location. In the remaining part of this chapter, instead of the official company names, the following names are used: company A, company B, and company C. Figure 1 shows that mostly men between the ages of 25 and 45 are employed in all three companies. The selected companies' headquarters are all located in the city center and are well connected to the network of bicycle paths [14].

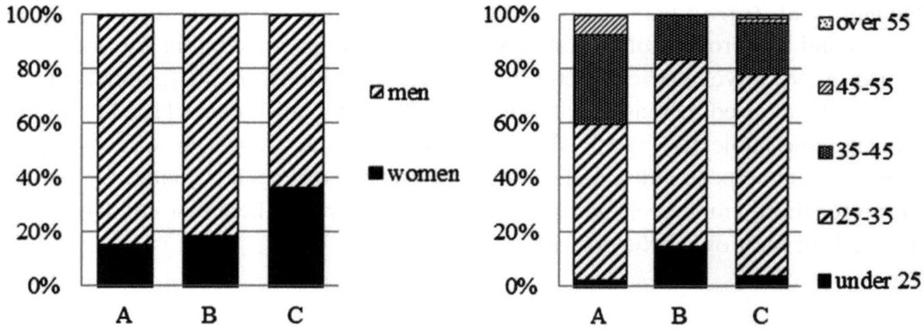

Fig. 1 Gender and age structure of the surveyed corporate entities (source: information received from the surveys)

Although the size of particular companies varies (respectively, A-150, B-80, C-240 employees), all entities are classified as medium-sized enterprises. Due to the size of the divisions it is possible to obtain a representative sample size and provide statistically significant results.

Survey Data Collection

An online questionnaire was identified as an appropriate method for data collection among employees of the IT sector. The survey was developed and conducted through the web service ankieta.pl. An invitation to participate in the survey, along with a link to the questionnaire, was sent by e-mail to each and every employee of the selected companies. The questionnaire consists of eight questions, including three closed-ended questions, one completely unstructured open-ended question, two matrix questions, and two demographic questions. The estimated time for completion of the questionnaire is 10 min.

Survey Results

In company A, 97 people filled out the questionnaire, within 9 days. This represents 65 % of all employees. In company B 48 individuals filled out the questionnaire within 21 days, which represents 60 % of all employees. In company C, the data collection period lasted 17 days, and 108 employees replied, which represents 45 % of the staff.

Closed-ended questions covered the following issues: commuting distance, bicycle availability, and bicycle commuting frequency. Most employees of each of the three companies live at a distance of 2–15 km from the premises of the company (Fig. 2a). The range of distance that is attractive for the bicycle commuters is

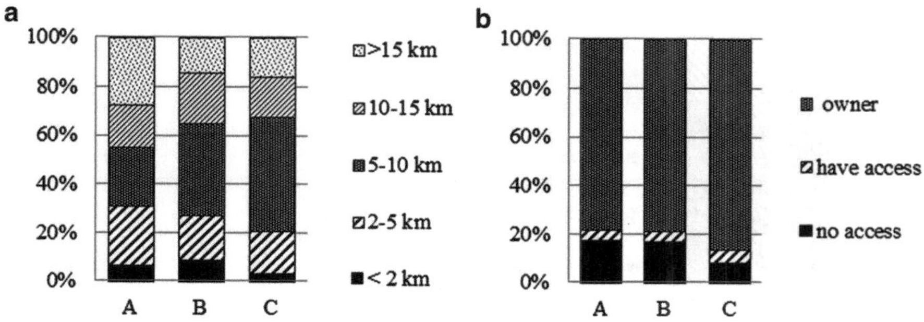

Fig. 2 (**a**) Distance to work, (**b**) availability of bicycle among employees of individual companies (source: information received from the surveys)

Fig. 3 Frequency of commuting by bicycle among employees of individual companies (source: information received from the surveys)

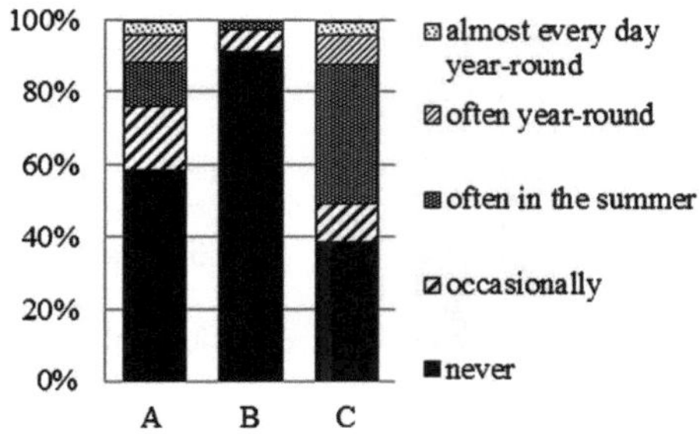

estimated between 1 and 15 km [15]. The employees' bicycle availability is also similar in the surveyed companies. The average of 86 % of the employees have the possibility to use a bicycle, including the average of more than 81 % of bicycle owners (Fig. 2b). The convergence between the companies at bicycle availability and commuting distance limits the impact of those factors on the results.

Answers to the question about the frequency of commuting by bicycle differ significantly between the companies (Fig. 3). A comparison of the number of people who never use the bicycle for commuting shows the following results: company A—59 %, company B—92 %, and company C—39 %. In general, in company B, bicycle is not used as a means of commuting. In companies A and C, the bicycle has a significant share in the modal split.

Convergent answers were also given to a few of the matrix questions about the factors affecting the decision on whether to use the bicycle as a commuting mode. Weather conditions are important for the majority of the employees in all companies (Fig. 4a). Weather is regularly mentioned in bicycle surveys as an important factor in the decision-making process. These data correspond to the comparison of urban data on the number of cyclists [16] to historical weather data [17]. Seasonal variations do not play a role as important as daily–short run variations [8]. Besides,

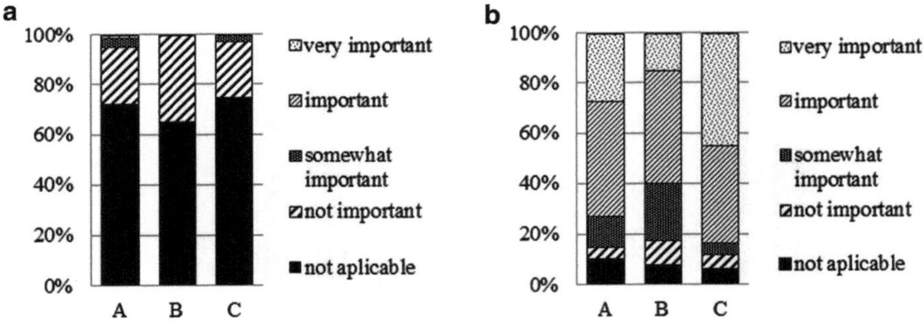

Fig. 4 The impact of (**a**) superior opinion and (**b**) weather on employees' decisions of not commuting by bicycle (source: information received from the surveys)

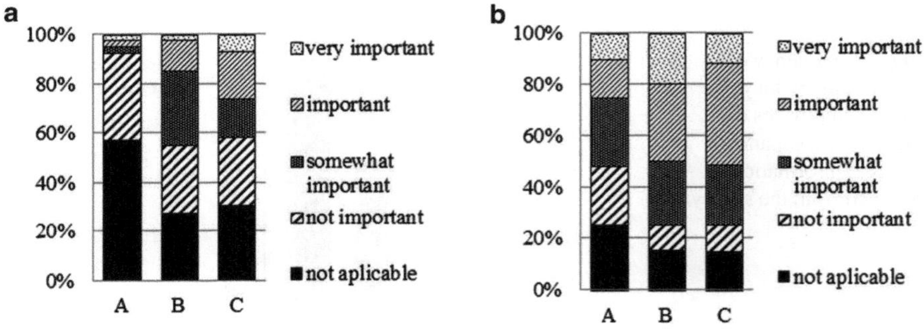

Fig. 5 The impact of (**a**) the fear of theft and (**b**) the need for a change of clothes after biking on employees' decisions of not commuting by bicycle (source: information received from the surveys)

in none of the studied cases, the opinion of a superior on the bicycle commuting to work matters (Fig. 4b).

The fear of bicycle theft is comparatively higher in company B and company C. Almost 60 % of the employees indicated that it as an important deterrent from commuting by bicycle while in company A most employees stated that this aspect does not matter at all (Fig. 5a). For employees of companies B and C the necessity of changing clothes after bicycle riding is a significant barrier in using the bicycle. Among the employees of company A this problem is not that important (Fig. 5b).

The fear of exhaustion or sweating is an important criterion which limits the propensity to commute by bicycle for the majority of the employees in all three companies (Fig. 6a). Some deviation in the answers occurs for the company C when determining the impact of travel time for a decision on bicycle commuting. In companies A and B, the average of 70 % of the respondents declare that the travel time is a factor taken into account when choosing means of transport and for more than half of the respondents' time is an essential or a very important factor in the selection. Meanwhile, in company C it is significant for no more than 30 % of the employees (Fig. 6b).

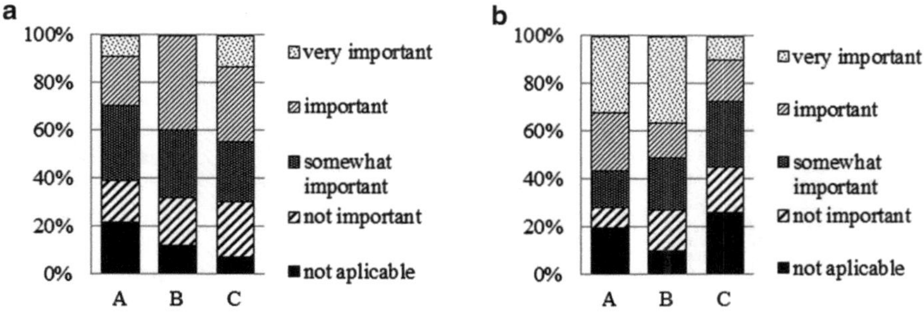

Fig. 6 The impact of (**a**) the fear of fatigue or perspiration (**b**) of fact that bicycle commuting time is longer than other means of transport on employees' decisions of not commuting by bicycle (source: information received from the surveys)

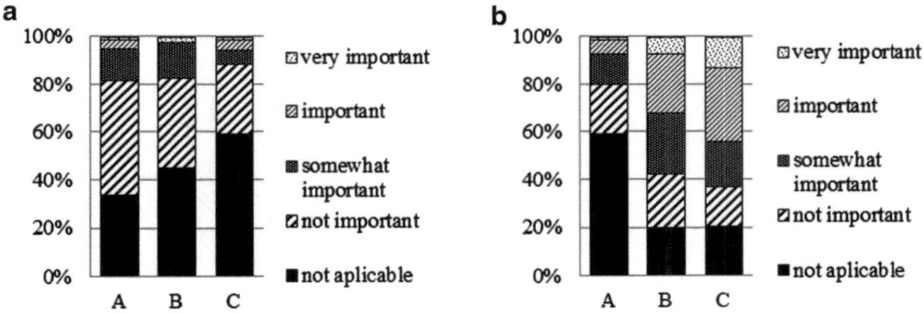

Fig. 7 The impact of (**a**) the need to bring the bike up the stairs and (**b**) dress code on employees' decisions of not commuting by bicycle (source: information received from the surveys)

The answers to the question about the barriers in the workplace were quite similar in all companies (Fig. 7a). Employees did not indicate physical barriers in the workplace. There is a significant deviation of responses among the employees of the company A to the question about the dress code (Fig. 7b). In the company A the required clothes do not affect the decision on the use of bicycle while commuting while in companies B and C it is found as an obstacle.

Responses to the question on cycling facilities in the workplace showed significant differences (Fig. 8a). Employees of companies A and C almost unanimously confirmed the ability to store a bicycle safely; moreover half of the respondents have used these facilities. In company B opinions are divided as to the occurrence of the safe bicycle parking and only 10 % have been using them occasionally. Even higher deviation was seen in the responses to the question on whether there is a shower in the workplace (Fig. 8b). In company A the staff confirmed the shower existence and even 30 % have used it with varied frequency. In company B the respondents almost unanimously confirmed the lack of shower facilities. On the basis of the responses provided by employees of the company C it is difficult to determine whether the shower actually is located in the office. Almost 6 % have no knowledge on the matter, and 64 % claim that there is no shower in their workplace,

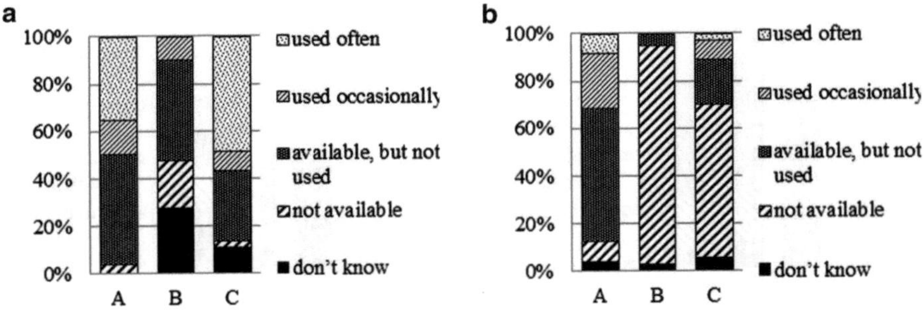

Fig. 8 Employees' awareness about the existence in the workplace and use of (**a**) a safe place to store the bike or (**b**) a shower (source: information received from the surveys)

Fig. 9 Employees' awareness of the existence in the workplace a place to change the clothes (source: information received from the surveys)

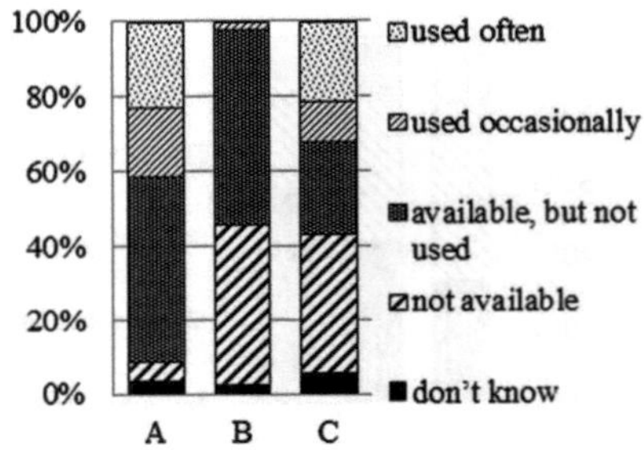

18 % claim that there is one but they have not used it, 8 % declared that they occasionally use it, and 2 % declared to use it frequently.

The company A employees also unanimously confirmed that the company offers a changing room for bicyclers (Fig. 9). Almost 42 % of the employees have used this facility. While 50 % of the employees of company B do not find the place to change, the rest indicated the existence of such a place, but admitted to not having used it. This discrepancy may occur due to a slight ambiguity within the question as it concerned a place to change clothing rather than a dressing room as such. In the company C the answers are divergent. Almost half of employees do not confirm the existence of the changing room when the rest confirmed, as in company B. In contrast to the company B over 11 % of the respondents have used a changing room occasionally and 22 % frequently.

Responses to the **open-ended question**, about factors influencing the decision not to use the bicycle for commuting, other than those mentioned in the questionnaire, can be grouped into the following major categories:

- The necessity of a car for errands before or after work
- Personal pro-automotive preferences and habits
- Dissatisfaction with transport network [14]

Conclusions and Applications

The initial assumptions were confirmed by the results of the research as regards the study of the group. The third element of the decision-making process—an analysis of destination barriers—displayed a crucial importance when considering bicycle commuting. Among many other work-related factors, those specific to the lower levels of Schein's organizational culture have proved to be crucial. As it can be seen in the survey results of the **company C**, the discrepancy occurs between specific levels as seen in Schein's model. At the level of artifacts: although there is a place to store the bicycle, there is no dressing room, so employees use other rooms to change clothes. If there is a shower then few are aware of this. The casual business dress code operates in the office. So for the part of the respondents having to change clothes is a deterrent. At the level of values and assumptions: commuting by bicycle is generally acceptable, no one is afraid of the lack of approval from the supervisor, and the team is young and active. Despite the lack of dedicated facilities, the percentage of cyclists is the largest among the surveyed companies.

In the **company A** the organizational culture is coherent. There are facilities for bicyclers such as changing rooms, showers, and parking places. To recognize the level of values and assumptions it was necessary to conduct additional personal interviews. Employees confirm acceptance of the company authorities for commuting to work by bicycle. The employer is responding to the growing needs of cyclists and, whenever it is possible, offers solutions for their convenience. As expected there is a high share of cyclists among employees.

In the **company B** organizational culture is coherent as well. There are no facilities for bicycler and commuting by bicycle is perceived as inappropriate. Existing rigid dress code reduces the propensity to commute by bicycle. Consequently, the bicycle share in modal split is negligible.

Bicycles are not used as extensively as they could be. The outcome of this research fosters a new approach in maximizing such a mode shift. Organizational culture shapes the culture of the environment through the transfer and promotion of standards adopted in the organization. The transport patterns propagated in the workplace affect directly the employees' transport choices and indirectly the wider population as a result of diffusion of values and norms. In addition, urban traffic is the sum of the movements of urban communities in which subgroups can be distinguished, among other, to the purpose of travel and employment. The most effective tool for managing employees' mobility is mobility plans. Part of the company's mobility plan is dedicated to promoting and motivating to active mobility. The outcome of the research indicates the need for increased emphasis to invisible layer of organizational culture in order to ensure greater effectiveness of the mobility plans.

References

1. Green Paper: Towards a new culture for urban mobility, COM/2007/0551 final.
2. Fernández-Heredia, A., Monzón, A., Jara-Díaz, S.: Understanding Cyclists' perceptions, keys for a successful bicycle promotion. In: Transportation Research Part A: Policy and Practice, vol.63, pp. 1–11. Elsevier (2014).
3. Baltes, M. R.: Factors influencing nondiscretionary work trips by bicycle determined from 1990 U.S. census metropolitan statistical area data. In: Transportation Research Record No. 1538, pp. 96–101, TRB, National Research Council, Washington D.C. (1996).
4. Sierpiński, G.: Zachowania komunikacyjne osób podróżujących a wybór środka transportu w mieście, Prace naukowe Politechniki Warszawskiej, Transport, vol. 84, pp. 93–106.Warszawa (2012).
5. Case Study No.1: Reasons Why Bicycling And Walking Are And Are Not Being Used More Extensively As Travels Modes, FHWA, U.S. Department of Transportation (1992).
6. Heinen, E., Maat, K., Wee, B.: The impact of work-related factors on levels of bicycle commuting, Paper submitted for presentation at the 88th Annual Meeting of the Transportation Research Board, Washington D.C. (2009).
7. Wardman, M., Tight, M., Page, M.: Factors influencing the propensity to cycle to work. In: Transportation Research Part A: Policy and Practice, vol. 41(4), pp. 339–350, Elsevier (2007).
8. Stinson, M. A., Bhat, C. R.: Frequency of Bicycle Commuting: Internet-Based Survey Analysis. In: Transportation Research Record. Pedestrians and Bicycles; Developing Countries. pp. 122-130. Transportation Research Board (2004).
9. Kamargianni, M., Polydoropoulou, A.: Development of a Hybrid Choice Model to Investigate the Effects of Teenagers' Attitudes Towards Walking and Cycling on Mode Choice Behavior. In: Transportation Research Record: Journal of the Transportation Research Board, Travel Behavior, vol. 1, pp. 151–161.Transportation Research Board (2013).
10. Anable, J.: Complacent car addicts or aspiring environmentalists? Identifying travel behaviour segments using attitude theory. In: Transport Policy, vol. 12, pp. 65–78. (2005).
11. Heinen, E., Maat, K., Wee, B.: The role of attitudes toward characteristics of bicycle commuting on the choice to cycle to work over various distances. In: Transportation Research Part D: Transport and Environment, vol. 16, pp.102–109. Elsevier (2011).
12. Bywalec C.: Ekonomika i finanse gospodarstw domowych, Wydawnictwo Naukowe PWN, Warszawa (2009).
13. Czerska M.: Zmiana kulturowa organizacji - wyzwanie dla współczesnego menedżera, Wydawnictwo Difin, Warszawa (2003).
14. Okraszewska, R.: The impact of organizational culture on employees' decisions on the selection of bicycles as a form of transportation to work, in the context of creating a new culture for urban mobility, based on an example of three IT companies located in Gdansk. In: Logistyka, vol.4, pp. 3101-3110. Instytut Logistyki i Magazynowania (2014).
15. Petritsch, T. A., Landis, B. W., McLeod, P. S., Huang, H. F., Scott, D.: Energy Savings from Provision of Bicycle Facilities. In: TRB 87th Annual Meeting Compendium of Papers DVD, Washington D.C. (2008).
16. Urban bike counters, http://www.rowerowygdansk.pl/start,169,67.html, view date 26.03.2015.
17. Archival weather forecast, http://pogoda.interia.pl/archiwum-pogody, view date 26.03.2015.

Risk Factor Classification GEMIO in the Planning Phase of Logistic Project Management

Dorota Książkiewicz

Abstract Logistic project management consists of a few phases, which may be generally defined as vision, planning, implementation, control, and project closure. In projects which involve logistic planning, logistic chain management, and physical distribution of goods, risk management issues are supposed to be included. Moreover they are often critical for the safety and security of the project implementation. Precise definition of risk factors is often difficult; however it is essential in the process of logistic project planning. Risk factors are individual for each logistic chain. They should be classified and their significance for the project should be assigned. Risk factors' classification GEMIO is a tool, which can be used to define risks in logistic processes and to plan the necessary safety and security procedures.

Keywords Risk management • Risk factors • GEMIO classification • Project management • Logistic planning

Introduction

Contemporary logistics has been shaped by globalization processes. Goods are moved around the globe and new technologies are used to ensure that the logistic processes are safe and efficient. When it comes to certain goods and logistic chains, project management is a useful tool to thoroughly plan, implement, and control the processes. Risk management is an important issue that influences logistic project management at all stages. The variety and specialization of logistic processes as well as the changes that constantly appear in the economic, geographical, and political surroundings of logistic companies are the reasons why logistics is one of the most risk-prone activities. The more complicated the logistic chain, the more risk factors should be considered. The nature of risk factors that occur in logistic

D. Książkiewicz (✉)
Transportation Policy Department, University of Gdańsk,
Jana Bażyńskiego 1A, 80-952 Gdańsk, Poland
e-mail: d.ksiazkiewicz@ug.edu.pl

M. Bąk (ed.), *Transport Development Challenges in the Twenty-First Century*,
Springer Proceedings in Business and Economics,
DOI 10.1007/978-3-319-26848-4_19

chain may vary due to the scope of the logistic chain (global, national, regional, local), the number and characteristics of the companies and institutions involved, and also the modes of transport and types of vehicles and machines used. Some risk factors are directly connected with the type of goods included in the project operations. Especially, when it comes to international transport processes, often intermodal, various types of risk appear. It is essential for logistic managers to define the risks at every stage, describe the possible consequences, and prepare the necessary safety procedures and resources. Yet the most difficult part of logistic risk planning is to foresee and define the risk factors due to the planned logistic chain. An additional difficulty within risk assessment process is that the nature of risk factors changes over time as new forms of risk, such as terrorism, pandemic flu, and the recent economic collapse within the financial sector, along with the consequential global economic crisis, serve to illustrate the borderless nature of risk [1].

Logistic Project Management

The first step to logistic project management consists of the visualization of processes. At this phase the main goal of the project is defined and the expected results are listed. Besides the tools and methods are set. The basic shape of the logistic chain is defined. Possible problems in implementation of the project are described. The second phase consists of detailed planning. Possible business partners are identified and assessed. The responsibility for certain tasks is divided and assigned to selected partners. Technical, organizational, and financial details of the project are prepared, along with the plan of necessary activities. At this point risk factors have to be predicted, according to expected problems and external economic, geographical, and political conditions. Their importance and impact on the logistic chain need to be evaluated. Safety procedures in case of the certain risk are established and responsibility for safety tasks is assigned. Detailed planning allows for effective implementation of the project and it helps to reduce operational risk. In the implementation phase effective information transfer is of great importance. In relation to physical distribution of goods, the issues of security and safety supervision become important. Other quality aspects, such as timing, precision, and effectiveness of warehousing and delivery, need to be constantly checked. Control phase allows for checking whether any corrections need to be done, according to the goal or the methods and tools that are used. Control procedures planned in the second phase are implemented, and they usually consist of quality control concerning goods and/or processes on each part of the logistic chain. In the last phase the results are analyzed. Economic and organizational analysis shows the most and the least effective processes and allows to identify the weak points of the logistic chain, which can be then modified in the launch of the following projects. Table 1 shows the phases of logistic project management and the main goals and activities suitable for each phase.

Logistic projects may include various logistic services, i.e.: supply and production, quality control, transportation, warehousing and terminal operations, value-added

Table 1 Phases and activities in logistic project management [2]

Phase	Goals	Activities
Vision	– Main goal of the project – Expected results – Basic shape of the logistic chain – Preferred methods and tools	– Economic analysis – Organizational possibility – Problem definition – Description of methods and tools
Planning	– Responsibility shared between partners – Risk factors and procedures – Acquiring resources (financial, technical, infrastructural, etc.)	– Identification and assessment of business partners – Identification of risk factors and risk evaluation – Description of safety procedures – Contract signing for assigned tasks
Implementation	– Ensuring safe and efficient delivery and/or storage or value-added process	– Information transfer – Schedule control – Security and safety supervision
Control	– Control procedures – Control planning versus implementation	– Constant monitoring of goods – Quality control
Closure	– Economic and organizational analysis of the results – Conclusions	– Cost analysis – Effectivity measurement – Experience gained, useful for future projects – Insure staff awareness and trainings

processes, distribution, sales, and reverse logistics. They may include the management of complete logistic chains or just some of its parts and processes.

Risk Management in Logistic Projects

We now seek our resources across large distances with the result that we create new forms of potential vulnerabilities in terms of our supply chains and the continuity of supply of those resources [3]. Risk management therefore influences logistic project at every stage. Firstly, the visualization and the goal setting have to take into account whether the project goal is possible to accomplish in terms of safety and security conditions. For example, a local military conflict in the destination area may make it impossible to accomplish the delivery of goods. However, the condition may not be final; sometimes it is possible to avoid the conflict area by choosing an alternative delivery route and/or transport mode, in which case it is essential to evaluate the costs and risk.

The planning stage of logistic project management is the stage, when most risk management work needs to be done. It allows to prepare for probable but also for those unexpected difficulties that may occur during the physical goods flow and the information flow within the project. Logistic processes are precisely analyzed in terms of risk factors and their possible impact. The most important issues of risk analysis are:

- Geographic, economic, and political situation on the route of delivery
- Transport and logistic infrastructure on the route and in reloading points
- Characteristic of the goods
- Climate and weather conditions
- Safe and efficient information flow

The issues mentioned above remain general risk groups, of which unique risk factors for every logistic chain should be derived. However, logistic managers often find it difficult to indicate the accurate risk factors in relation to planned logistic operations. When risk factors are indicated, their importance and possible consequences need to be addressed. Thus, the security procedures are described along with issues such as:

- Goods preparation and packaging standards
- Expected level of insurance protection
- Expected staff qualifications
- Setup of a crisis management unit
- Communication technologies and rules

At this point, safety and security procedures need to be planned with as much precision as possible, concerning each unique logistic process. Together with the procedures, control rules must be set—complying with standards will be checked regularly according to plan during the implementation and control phase. As a result of control, modification of operation rules is often needed to ensure better quality of the logistic processes. At the closing stage of the project, the results are summed up and compared to initial requirements. The risk analysis at this point concerns the following issues:

- Which of the prepared risk procedures were actually used
- Degree to which the procedures were effective
- Necessary modifications to risk management strategy
- Other issues that require security and safety supervision in future projects

Therefore risk management influences logistic projects by setting and executing security standards in logistic operations.

Recognition of Risk Factors According to GEMIO Classification

Detailed risk factors' analysis takes place when the shape and specification of logistic chain are already planned. In order to classify the variety of risk factors influencing the level of safety in logistics, all processes within the logistic project should be acknowledged and described. The processes are then subject to the integrated risk management approach, in which risk identification remains the basic activity. According to ISO 31000 (2009) standards, the aim of risk identification is "to generate a comprehensive list of risks based on those events that might create, enhance, prevent, degrade, accelerate or delay the achievement of objectives. It is important to identify the risks associated with not pursuing an opportunity. Comprehensive identification is critical, because a risk that is not identified at this stage will not be included in further analysis" [4]. GEMIO classification is one of the tools to analyze a logistic project in relation to risk identification and evaluation. The name GEMIO stands for the words describing main risk-generating areas in logistic processes. These potential risk-generating areas are [5]:

- *Geopolitics*—includes risk factors of a geographical or political nature, such as military conflicts, terrorism, piracy
- *Environment*—includes risk factors resulting from natural environment, such as earthquakes, floods, hurricanes
- *Market*—consists of risks resulting from difficult or variable market conditions, for example the necessity to cooperate, raising quality of processes, variability of fuel prices, severe market competition
- *Infrastructure*—includes risks concerning availability of technical conditions, vehicles, machines and qualified human resources, their quality and quantity required at each stage of the logistic chain
- *Organization*—risk factors resulting from the process of preparation and implementation of logistic procedures, human mistakes, documentation failures, etc.

There are risk factors generated in all mentioned areas in each supply chain. However, their importance and the level of consequences may vary according to the specific organizational features of the supply chain. Moreover the risk-generating areas may in some cases become connected, and as a result risk factors may be escalated or declined. Possible interactions between GEMIO areas are shown in Fig. 1.

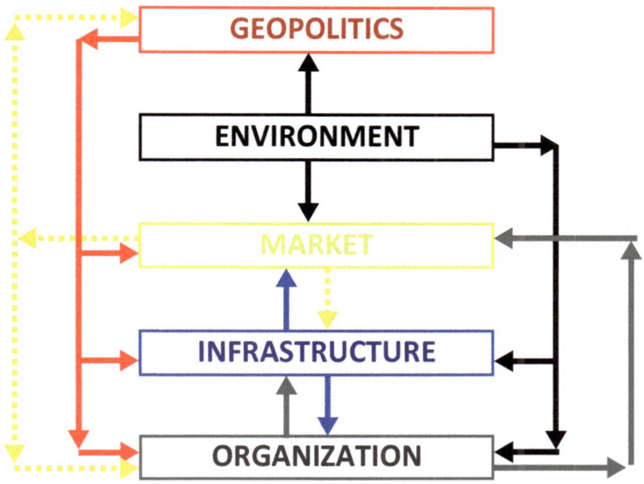

Fig. 1 Possible interactions between GEMIO risk areas [2]

Market is considered to be the central risk-generating area, where most of the dangerous occurrences make an impact. Moreover, market area itself is a source of numerous risk factors. However market must not be the only field subject to risk management within a logistic project. Market risk appears mainly due to changing market conditions; for example when law regulations change, companies may have to adjust to new competition rules, comply with additional or higher standards, etc. Economic collapse on the global financial market, as occurred in 2009, usually results in many risk factors for logistics, such as decreasing cargo volumes that leave transport vehicles and warehouse supply inefficient, and logistic operational revenues decline because of variations in exchange rates. Many risk factors also result from economic uncertainty. Logistic project management requires acknowledging an operational budget for the time of project duration, which can be considerably affected by changing economic conditions, for example a change in the level of marine freight prices. Therefore at the planning phase of the logistic project, the stability of market surroundings should be assessed, and suppliers and other partners involved in the supply chain should be checked in terms of risk within:

- Their past and recent activity on the market
- Their competitive strategy
- Contract safety conditions
- Their financial situation
- Their professional reputation

Market competition may also be a source of risk for logistic operations, resulting for example in suppliers breaking up the contract or other logistic companies gaining competitive advantage by using new technology.

Global and international supply chains may be affected by market risk in every geographical region they cover. Religious beliefs, cultural differences, and language problems may in many ways be a source of problems and hazards and should be acknowledged in the risk identification process.

The area of market risk remains in relation to the other risk field—*organization*. A number of risk factors are derived from inside the organizations and they may vary depending on the shape of the organization structure, motivation system, communication system, the level of staff qualifications and experience, and their ability to communicate and cooperate within the organization and with external partners as well. In this group of risk factors the most common are:

- Delays and mistakes caused by misunderstandings and other communication problems
- Human mistakes, for example concerning transport and customs clearance documentation
- Lack of compliance with standards, either as the result of lack of knowledge or unwillingness to comply
- Lack of supervision over critical tasks and activities

Communication issues between all the parties involved in a project are one of the most risk-generating factors. Logistic supply chains often comprise of many different enterprises and institutions; therefore the coordination of activity requires setting up of an effective and reliable communication system.

Infrastructure is the next important field that requires thorough examination in logistic projects. Sufficient and reliable transport and logistic infrastructure is one of the preconditions for logistic planning. Technical equipment, vehicles, storage places, terminals, etc. determine the quality of logistic processes. Availability of infrastructure must be analyzed in relation to each and every step in logistic supply chain, as it often becomes a potential risk factor, especially in projects of wide geographical range. Terminal charges and prices for infrastructure use may also affect the possibility to conduct certain logistic operations.

Environment is often considered to be the most obvious, yet unpredictable risk factors' area. A wide spectrum of weather hazards has to be taken under consideration, regarding a geographical region in which logistic operations are planned to be conducted. The most common risk factors in this area remain:

- Supply delays resulting from bad weather conditions
- Accidents caused by harsh weather conditions
- Cargo loss or damage
- Logistic infrastructure damage
- Additional costs of storage at times of delay in terminal or transport operations

Some of weather-caused hazards may be locally predictable; however their intensity, range, and effect remain usually unknown.

Geopolitics, as an area of potential risk in relation to logistic processes, comprises of all political, economic, and social occurrences which, in a certain period of time, may affect the shape and implementation of logistic processes. Among these risks, the most common are:

- Military or economic conflicts which result in closing markets, trade barriers, or destruction of logistic infrastructure
- Acts of terrorism or piracy
- Sudden changes in security procedures, esp. in marine ports and terminals
- Launch of local freight war surcharge by shipping lines
- The necessity to switch to safe transport routes, resulting in supply chain reconfiguration and possible longer delivery times

Geopolitical risks may to certain degree be predicted; nevertheless their impact on logistic costs is usually substantial. For example in 2015 expanding military conflict disabled marine trade through marine ports of Libya, the fight for power in Yemen has serious implications for the region as well as for international marine trade, and the escalation of piracy incidents in the Gulf of Guinea causes turbulences in marine transport. The consequences of such risks may be long lasting.

As shown in Fig. 2, GEMIO classification is useful at all stages of logistic project management, although it is mostly planning and implementation phase that can benefit from a detailed recognition of risk factors.

At the closing stage of the project, risk management experience is summed up and market, infrastructural, and organizational conclusions are usually formed to help increase effectiveness in future project planning.

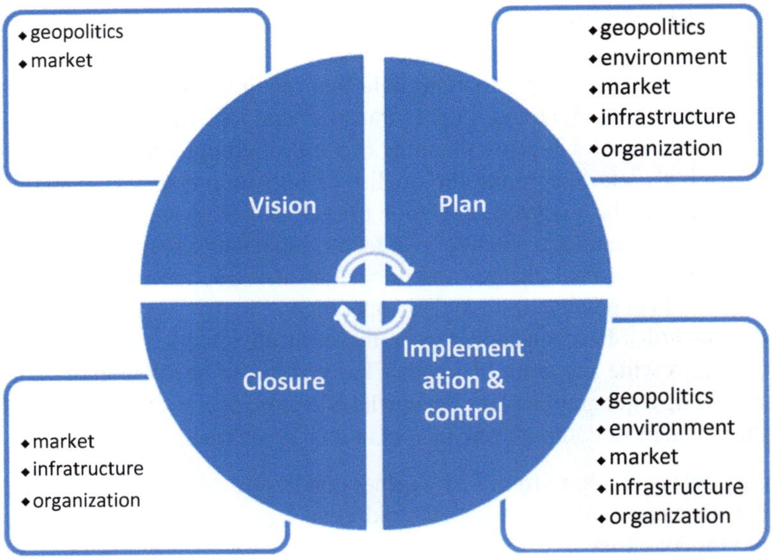

Fig. 2 GEMIO classification areas considered in subsequent stages of logistic project planning [2]

Risk Assessment and Prevention

GEMIO classification proves to be effective in listing possible threats to logistic operations. It helps to recognize risk covering all possible operation work areas. Risk identification should be as detailed as possible; thus each of the identified risk factors should be described according to such issues as:

- The type of force or event causing danger
- Probability of appearance
- Consequences for logistic operations
- Additional logistic costs
- Expected durance
- Prevention tools, safety procedures
- Crisis-handling methods, ways to minimize crisis impact on the project

In every logistic project there is an acceptable level of risk that can be handled to every risk factor. There are cases in which the probability of appearance of a certain risk, the range of its consequences, or expected additional operational costs go beyond the acceptable level. As a result, project completion is endangered to a degree that requires strategic decisions, either substantial reconfiguration of the supply chain to minimize risk or project shutdown.

When the risk to which the project is exposed is acceptable, prevention tools and safety procedures need to be prepared and followed. Prevention tools are most effective in relation to risks derived from the field of *organization*; they would be difficult to establish and much less effective according to *market* and *infrastructure* fields, whereas in case of *environment* and *geopolitical* factors it is often impossible to prepare prevention altogether and safety procedure use is limited to establishing crisis headquarters together with responsibility assignment within crisis-handling methods. The most common prevention tools are:

- Staff trainings, especially according to emergency procedures
- Quality control in relation to goods at every stage of the logistic chain as well as in relation to machines, vehicles, and infrastructure
- Regular update and control of information and security systems
- Establishing emergency reaction units together with other business partners in the supply chain

Emergency procedures may be established on the basis of ISO 31000 standards; however they should refer specifically to every recognized risk factor and should consider the specific features of planned logistic activity. Emergency procedures should be at all times consulted and may be established in cooperation with other subjects in the supply chain.

Summary

Risk management is an essential part of logistic project management, especially in relation to the evolving nature of risk on global market. The more complicated the logistic chain, the greater the set of risk factors that has to be recognized and considered. Global supply chains result in the time-consuming and geographically wide transport routes which is a risk-expanding feature for logistic projects as well. The basic objective of risk management is to minimize the risk in logistic operations and to establish efficient emergency tools and methods for crisis handling. It is impossible to eliminate risk altogether, but it can be reasonably assessed and handled contributing to the overall success of the project.

References

1. Smith D., Fischbacher M: *The changing nature of risk and risk management: The challenge of borders, uncertainty and resilience*. Risk Management (2009) 11.
2. Author's own study.
3. Fischbacher-Smith D: *Destructive landscapes – (Re)framing elements of risk?* Risk Management (2011) 13.
4. Borghesi A., Gaudenzi B: *Risk Management*. http://link.springer.com/search?query=Risk+Identification++Basic+Stage+in+Risk+Management#page-1.
5. Książkiewicz D., Mierkiewicz D: *Bezpieczeństwo we współczesnych łańcuchach logistycznych*. (w:) "Współczesne problemy rozwoju transportu zrównoważonego". Praca zbiorowa pod red. R. Rolbieckiego, Zeszyty Naukowe Uniwersytetu Gdańskiego, Ekonomika Transportu i Logistyka, nr 43. Wydawnictwo UG, Gdańsk 2012.

Use of Economic and Econometric Analysis in the Financial Standing Diagnosis of Haulage Enterprises, Case Study: Trans Polonia S.A.

Andrzej Letkiewicz and Michał Suchanek

Abstract This chapter focuses on the specifics of financial standing in the haulage enterprises. The depreciation cost is the main factor that creates the financial result in such enterprises. While decreasing the value of the net profit, the depreciation cost is strictly correlated with property, plant and equipment of the enterprise and thus with the sales revenue. The analysis uses both the classic method of economic analysis and econometric methods, such as regression analysis, so that it may both facilitate monitoring the behaviour of financial standing of the enterprises and be considered a factor in the decision-making, especially in the investment policy.

Keywords Road transport • Haulage enterprises • Motor vehicle depreciation cost • Regression analysis in transport enterprises

Introduction

Enterprises come into existence so as to multiply their market value, thus multiplying the value of their owners' paid-in capital. This is a long-term objective. In the short term it decomposes into a spectrum of short-term objectives, out of which, maximising the yearly profit is the most important one. The application of a methodology for profit calculation in accordance to the accounting and tax law means compiling the revenues with their corresponding costs. One of these costs is the depreciation cost which is specific due to the fact that its level is a result of law regulations on the one hand and the enterprise policy on the other hand. The yearly level of depreciation cost is a result of the value of owned assets and is therefore a natural consequence of previous investment expenditures and applied depreciation rates and methods. This situation creates a possibility to manage the performance of the company according to the goals, which it wishes to accomplish.

A. Letkiewicz (✉) • M. Suchanek
Faculty of Economics, Chair of Economics and Management of Transportation Companies, University of Gdansk, Jana Bażyńskiego 1A, 80-952 Gdańsk, Poland
e-mail: ekoalt@univ.gda.pl; m.suchanek@ug.edu.pl

M. Bąk (ed.), *Transport Development Challenges in the Twenty-First Century*, Springer Proceedings in Business and Economics,
DOI 10.1007/978-3-319-26848-4_20

Performance management through the appliance of depreciation control leads to a possibility of its increase or decrease. That allows the managers to create their relations with the owners (building a constant increasing profit trend leads to a positive evaluation of the management by the owners) and to control the level of financial accumulation which is the basis of long-term liabilities' repayment. This issues are of even higher importance in haulage enterprises which specialise in road transport due to the fact that they usually have a relatively low participation of non-transportation assets in total assets and they always have a possibility to obtain motor vehicles in a way which does not generate depreciation costs, that is, through the leasing contracts.

The role of depreciation in performance management and the information needs of corporate governance analysis led to the formulation of the chapter goal, which is to define the basic analysis conditions and characteristics of depreciation control on an enterprise. The additional goal is to construct an econometrics model which shows the effects of depreciation policy on basic enterprise parameters such as the value of tangible assets, revenues and the net profit on an enterprise. Financial statements of a haulage company, Trans-Polonia S.A., have been assumed as the basis for analytical data.

Revenues, Tangible Assets, Depreciation and Net Profit in a Road Transport Enterprise

Service sales revenues are of vast importance for a transport enterprise because they are a marker of social acceptance of service usefulness. On the other hand, they reflect the state of enterprise relations with the market. In consequence, the revenues, from the point of view of the enterprise, verify the perception of the enterprise services, the efficiency of the economic activity and the quality of management. From the financial point of view, the revenues are a source of the inflow of financial funds and are necessary to cover expenditures on tangible assets and the expenditures on created transport services. The level of revenues itself is of fairly low relevance. However, the comparison of the revenues with the level of expenditures on tangible asset purchases or manufacturing (which in the operational activity of the enterprise are identified through the depreciation cost together with other costs of service creation) is of high analytical value [1]. The freight haulage enterprise uses the system of conventional prices which is in essence an agreement to rent motor vehicles. Therefore, it can be stated that the level of revenues in such enterprises is an effect of their own carriage potential and the efficiency of executing orders.

Property, plant and equipment are the entirety of useful and complete assets which are supposed to be used in enterprises for a period of time longer than 1 year and whose value at the moment of purchase is higher than 3500 PLN [2]. Current assets, on the other hand, are materials, receivables and cash of the company. In a road transport company, the asset structure is formed mostly by PP & E. In the diagnosed enterprise, PP & E was 65.8 % of the overall value of assets. However, the current assets were comprised in 75 % of the receivables. Such a structure signifies that the group of current assets in an enterprise is of importance in the process of keeping liquidity, but not in the creation of revenues and economic efficiency. The latter are a direct result of carriage potential and technical background which assures the effectiveness of motor vehicles on hand.

The state of property, plant and equipment is a result of company investment policy. The level and structure of assets warrant the market potential of a haulage company—it is a basis of revenue creation, at the same time being a source of costs which have to be covered by the revenues if the company is supposed to generate net profit [3]. The depreciation cost is the cost marker of use of property, plant and equipment in the operational activity of the company. The enterprises are allowed to individualise the depreciation rates by increasing or decreasing them in comparison to the standard linear method. The applied rate of yearly depreciation directly influences the composition of two efficiency parameters—net profit and financial accumulation.

Net profit is a synthetic indicator of the results of enterprise economic activity. The pursuit to maximise profit is the main motivation for increasing the enterprise efficiency. Hence, net profit is determined by the ability to generate revenue and by the applied investment policy and cost policy, including the depreciation policy. The bottom line of the enterprise is also the basic criterion of decision-making and a basis for self-financing development of the enterprise. From the financial point of view, the net profit and depreciation are those factors which form the financial accumulation which is the basis for investment decision in the future. The financial accumulation, in turn, leads to the upkeep of value of property, plant and equipment.

Characteristics of the Enterprise Under Examination

Trans-Polonia S.A. is a specialised transport and shipping operator, whose economic activity mostly includes logistics of chemical liquids, liquid asphalts, liquid FMCG and controlled-temperature goods. The enterprise works across Europe and is managed from the headquarters in Tczew, Poland, and France. The customers of the enterprise are mostly the major petroleum companies. Cooperation on such a highly specialised market is based on the high quality of services, the reputation of the supplier and the ability to build mutual trust. Trans-Polonia S.A. meets industry standards such as ADR convention and ISO certificates. In recent years, its activity in the intermodal transport sector has grown, which combines different means of transport, using containers. Trans-Polonia S.A. organises intermodal transport within the country as well as on an international and intercontinental level. It is a public limited company, registered on the stock market in Warsaw since 2008. Financial standing, including the value of PP & E, yearly depreciation costs, revenues and net profits, has been shown in the table (Table 1).

Table 1 Financial standing of Trans-Polonia S.A

Year	Property, plant and equipment	Depreciation	Net profit	Sales revenue
2008	6905.34	669.00	254.80	13,811.00
2009	6766.00	813.00	500.00	12,899.00
2010	6472.00	845.00	1146.00	22,660.00
2011	13,394.00	1 097.00	5109.00	46,931.00
2012	20,153.00	2175.00	2092.00	49,308.00
2013	28,552.00	2699.00	2535.00	38,808.00
2014	38,630.00	4131.00	2616.00	65,286.00

In thousands PLN
Source: http://www.transpolonia.pl/pl/profil-gk-trans-polonia-sa

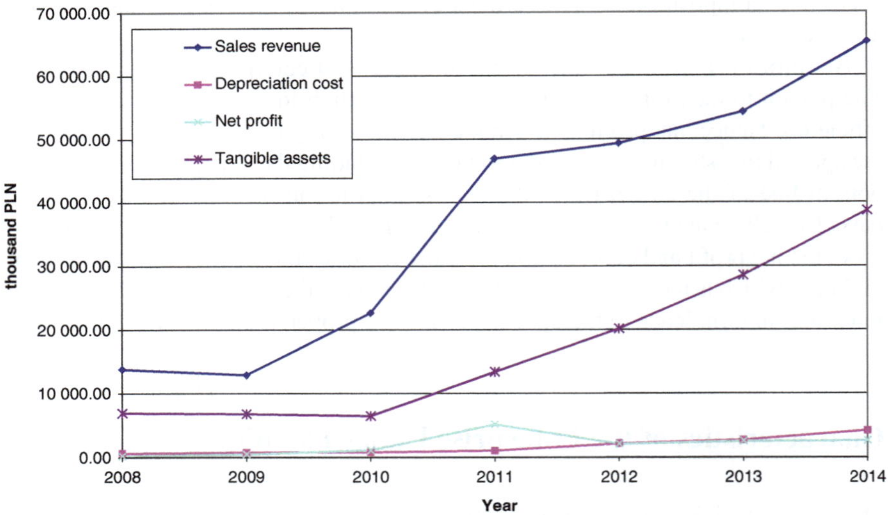

Fig. 1 Basic economic values in the enterprise (source: http://www.transpolonia.pl/pl/profil-gk-trans-polonia-sa)

The analysed period can be divided into two sub-periods which show the characteristics of the enterprise activity. In years 2008–2010, the value of PP & E was fairly stable, with a small tendency to decrease. However, in 2011, the value of tangible assets has doubled and developed a constant growth tendency up to a level of 38.6 million PLN in 2014. This happened due to large investment expenditures. The changes of the value of PP & E brought together changes in the level of depreciation costs, net profit and sales revenues as well. The levels and directions of changes of analysed values are shown in the figure (Fig. 1).

Method for Analysis and Result Analysis

The analysis of the impact of depreciation costs on the financial standing of the enterprise has been performed with the use of the following indicators (Table 2):

- Return on PP & E (net profit/PP & E × 100 %)
- Return on sales (net profit /sales revenue × 100 %)
- Financial accumulation (net profit + depreciation)
- Depreciation to sales coverage (depreciation/sales revenue)

During the analysed period, the depreciation cost has been calculated with the use of linear method throughout the whole usage period. The monthly costs begin at the moment the asset starts to be used and end at the moment the accumulated depreciation cost is equal to the initial value of the asset. Otherwise, they end if the asset is sold or removed from the financial statement—depending on what happens first. The depreciation rates in the enterprise depend on the group of tangible assets and are as follows [4]:

- Buildings 2.5–4.5 %
- Motor vehicles 7–20 %
- Machines and machinery 10–18 %
- IT hardware 10–30 %

Determining the participation of depreciation costs in sales revenue and financial accumulation is crucial for the analysis and prognosis of the enterprise financial standing. It is a result of the role of depreciation in the creation of financial result. When it comes to calculating the financial result, the depreciation cost is one of the factors which decrease the revenues; yet when it comes to the financial accumulation, the depreciation cost increases the value of cash on hand for the enterprise. Calculated values of the participation of depreciation in revenues and financial accumulation are shown in the table and the overall summary of the return on assets, return on sales and depreciation participations is shown in the chart (Table 3).

Table 2 Indicators of enterprise financial standing

Year	Return on PP & E (%)	Return on sales (%)	Depreciation to sales coverage (*n*-times)	Financial accumulation (in thousands PLN)
2008	3.69	1.84	20.64	923.80
2009	7.39	3.88	15.87	1313.00
2010	17.71	5.06	26.82	1991.00
2011	38.14	10.89	42.78	6206.00
2012	10.38	4.24	22.67	4267.00
2013	8.88	4.67	20.10	5234.00
2014	6.77	4.01	15.80	6747.00

Source: http://www.transpolonia.pl/pl/profil-gk-trans-polonia-sa

Table 3 Participation of depreciation in financial accumulation and in revenues

Year	Return on PP & E (%)	Return on sales (%)	Participation of depreciation in revenues (%)	Participation of depreciation in financial accumulation (%)
2008	3.69	1.84	4.84	72.42
2009	7.39	3.88	6.30	61.92
2010	17.71	5.06	3.73	42.44
2011	38.14	10.89	2.34	17.68
2012	10.38	4.24	4.41	50.97
2013	8.88	4.67	4.98	51.57
2014	6.77	4.01	6.33	61.23

Source: http://www.transpolonia.pl/pl/profil-gk-trans-polonia-sa

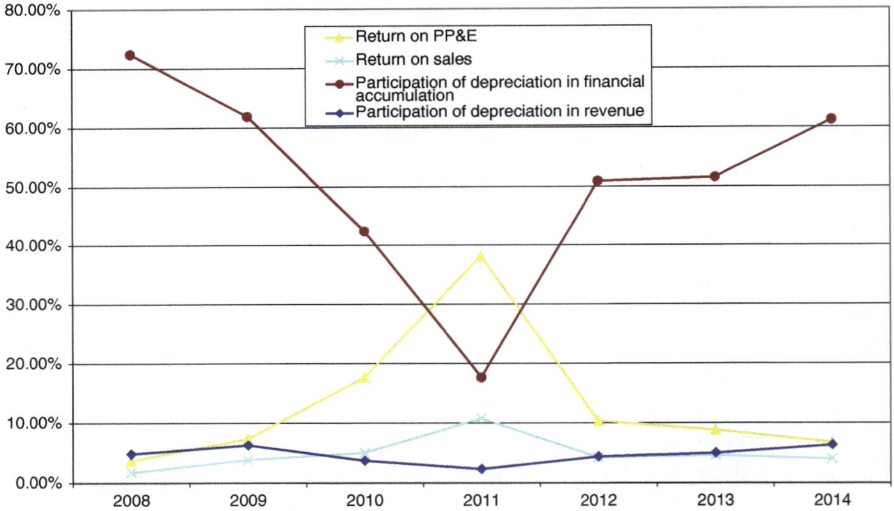

Fig. 2 Return on assets, return on sales and depreciation participations (source: http://www.trans-polonia.pl/pl/profil-gk-trans-polonia-sa)

The aforementioned characteristic confirms the result efficiency which the enterprise has achieved every year. However, these stats are also the basis for making future economic decisions. The analysis has shown that the year 2011 was groundbreaking for the enterprise. Function graphs shown in Fig. 2 confirm that feature. In 2011 the trend for all the parameters changed. In order to diagnose the strength of influence of the management policy on the situation, one has to perform a *y/y* dynamics analysis for all the parameters. The results for the return on sales, return on PP & E and the depreciation participations are shown in the graph (Fig. 3).

The course of dynamics characteristics shows that years 2011–2012 were crucial for the management of the enterprise as regards the decisions concerning the most important investment and exploitation parameters. Very favourable results in 2011 were put down in 2012. This led to a strong re-evaluation and resulted in a depreciation–revenues–accumulation balance in 2013–2014 that was similar to the one that happened earlier in 2009.

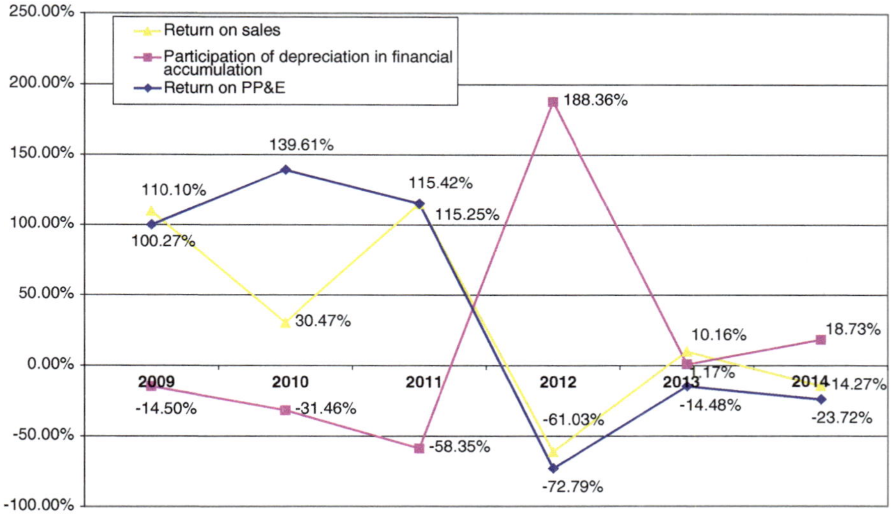

Fig. 3 Dynamics of the changes in financial accumulation, return on sales and return on PP & E (*y*/*y*) (source: http://www.transpolonia.pl/pl/profil-gk-trans-polonia-sa)

Table 4 Pearson correlation coefficients for the pairs of values

Variable 1	Variable 2	Pearson coefficient
Tangible assets	Depreciation	0.98
Tangible assets	Net profit	0.41
Sales revenue	Depreciation	0.86
Net profit	Depreciation	0.33

Source: Own estimates

Econometric Analysis

The econometric analysis begins with the analysis of correlation between the significant economic factors in the enterprise [5]—tangible assets, depreciation, sales revenue and net profit. Pearson correlation coefficient was calculated for the following pairs: tangible assets and depreciation, net profit and depreciation, tangible assets and net profit, sales revenue and depreciation. The results are shown in the table (Table 4).

From the economic point of view, these are all fairly high results. The value of the correlation between tangible assets and depreciation is a natural consequence of the applied depreciation method. However, an especially significant correlation between sales revenue and depreciation can be considered a proof of the importance of PP & E in the generation of sales revenue in the analysed enterprise.

The correlation analysis was a starting point for the regression analysis. Three models were calculated with the use of least square method:

$$\hat{\mathrm{NP}}_t = -0.1841\mathrm{TA}_t + 0.155\mathrm{SR}_t \quad (1)$$

$$\hat{FA}_t = 0.0001SR_t \quad (2)$$

$$\hat{NP}_t = 39811.8Dyn\_TA_t \quad (3)$$

Where:
NP_t = net profit in year *t*.
TA_t = tangible assets in year *t*.
SR_t = sales revenue in year *t*.
FA_t = financial accumulation in year *t*.
Dyn_TA_t = dynamics of tangible assets in year t ($TA_t/TA_{t-1} \times 100\ \%$).

The formal correctness of all the models was checked with the methods presented in Table 5.

Apart from these tests, all the variables were tested for their individual statistical significance using the *t*-student test and the coefficient of determination R^2 was calculated for all the models (0.82 for the first model, 0.91 for the second one and 0.92 for the third one). The models are formally correct and can thus be used as a basis for research.

The models altogether show a strong relation between the value of assets, depreciation costs, revenue and net profit. An increase of sales revenue by 1000 PLN leads to the increase of net profit by 150 PLN. An increase in the value of tangible assets by 1000 PLN leads to the decrease of net profit by 180 PLN, through the cost of depreciation. However an increase in depreciation should lead to an almost equally strong increase in sales revenue (strong Pearson correlation coefficient) which in turn leads to an increase in both the net profit and the financial accumulation.

Table 5 Formal testing of the models

Criterion	Statistical test	Null hypothesis	*p*-value model 1	*p*-value model 2	*p*-value model 3
Statistical significance of variables	F-test	The variables are collectively statistically non-important	0.02	0.0007	0.0005
Heteroskedasticity	White test	Residual variance of the variables in the regression model is constant	0.22	0.219	0.5226
Structural changes during the time period	Chow test	There are no structural changes (2011 assumed as the year of division)	0.85	0.86	0.42
Autocorrelation	Doornik-Hansen test (Lagrange multiplier)	There is no autocorrelation	0.34	0.43	0.93

Source: Own estimates

Hence, it can be said that the control of depreciation cost indirectly allows to control the net profit of the enterprise. Furthermore, the increase of value of tangible assets in the enterprise does lead to an increase in profit, as long as the dynamics of the increase are strong enough. This shows that only through an especially offensive investment policy can the company upkeep the acquisition of new orders and therefore the generation of net profit. If the haulage company ceases to aggressively extend its carriage potential it starts to implode and lose its market position. Therefore, managing to achieve a balance between newly purchased property, plant and equipment and the depreciation cost of the already used one can lead to an increase in net profit. In the case of Trans-Polonia S.A., an increase in the dynamics of the net value of tangible assets by one percentage point has led to an increase of net profit by 39,000 PLN.

Conclusions

The case study of the use of economic and regression analysis in the Trans-Polonia S.A. enterprises has allowed the authors to arrive at the following conclusions:

1. The depreciation cost is an extremely important factor in the road transportation enterprises and should be subject to particular control.
2. The depreciation cost in a haulage company is mostly the result of investment policy and depreciation policy on motor vehicles.
3. Factors such as net profit, sales revenue and depreciation cost can be analysed collectively with the use of economic analysis and econometric analysis.
4. The depreciation cost in a haulage company is directly correlated with the sales revenue and indirectly with net profit and financial accumulation.
5. The econometric regression analysis can be a valuable source of information for decision-making concerning the investments and depreciation policy in a haulage enterprise, as long as it is formally correct and stable.

References

1. Czubakowska K., Gabrusewicz W., Nowak E.: Przychody, koszty, wynik przedsiębiorstwa. PWE, Warszawa 2009.
2. Walińska E. (ed.), Rachunkowość finansowa, Oficyna Wydawnicza Wolters Kluwer Business, Warszawa 2010.
3. Berwick M., Farooq M., Truck Costing Model for Transportation Managers, Upper Great Plains Transportation Institute, 2003.
4. Consolidated Financial Statement of Trans-Polonia S.A. for 2014 http://www.transpolonia.pl/raport/396.
5. Cohen J., Statistical power analysis for the behavioral sciences, Lawrence Erlbaum Associates, New Jersey 1988.

Koleje Śląskie as an Example of H. Mintzberg's Structural Configuration Theory Application in Designing Organizational Structure of a Railway Operator

Robert Tomanek

Abstract In practice, the challenge of transport management lies in the development of organizational structures of enterprises—especially railway companies, which are usually characterized by a kind of hierarchical organizational structures. Meanwhile, rail companies in liberalized markets need to adapt to market conditions. The theoretical achievements in the field of management science and theory of enterprise can be utilized to achieve the abovementioned goals. One of the most interesting models based on practical research of organizational structures is the theory of structural configuration, developed by H. Mintzberg in 1979. This model was used to design the organizational structure of Koleje Śląskie Sp. z o.o. Mintzberg's model is based on five main parts of organization, five coordination mechanisms, nine design parameters, and four groups of conditions. It is a powerful diagnostic and planning tool. Mintzberg's model has proven to be useful in the evaluation and design changes to the organizational structure of Koleje Śląskie Sp. z o.o. The company was established as a so-called internal operator with the objective of providing passenger rail service in Silesia. It has been proven that H. Mintzberg's model can be successfully used in the creation of a flexible organizational structure adapted to the liberalization of transport markets. The proposed model was used in the construction of the organizational structure and organizational regulations that have been adopted by the company in 2014. The next step of the restructuring was a significant change in the corporate strategy of Koleje Śląskie Sp. z o.o. The aim of this chapter is to demonstrate the possibility of a successful application of the theoretical output of Henry Mintzberg in the field of railway transport.

Keywords Public transport • Railway transport organization • H. Mintzberg structural configuration theory

R. Tomanek (✉)
University of Economics in Katowice, 1 Maja 50, 40-287, Katowice, Poland
e-mail: tomanek@ue.katowice.pl

M. Bąk (ed.), *Transport Development Challenges in the Twenty-First Century*,
Springer Proceedings in Business and Economics,
DOI 10.1007/978-3-319-26848-4_21

Introduction

In the nineteenth century, railway companies became the birthplace of organizational studies, which was related to the necessity of such management of people and processes that would allow for carrying out complex projects, such as construction and operation of railway system. The contemporary railway transport operators also function in a complex market and sector space—namely, a system of specific market regulations, ownership relations, and technological conditions determined by the complexity of railway transport process. The organizational structure of a railway operator has to reflect this specificity, and moreover, it should be subject to ongoing audit and modifications enabling to adapt the company to changing operating conditions in a difficult environment.

The abovementioned conditions were ignored during the establishment of Koleje Śląskie Sp. z o.o. by the self-government of Silesian Province in 2010. The operator, who took over the provision of regional and local passenger railway transport services as a so-called internal operator, was established without taking reasonable organizational measures. As a result, the established company had non-transparent structures, and moreover, it lacked organizational units responsible for fundamental processes (e.g., related to rolling stock and finance management). The shortage of appropriate staff, deficiencies in rolling stock, and difficult weather conditions during the handover of duties from Przewozy Regionalne (a regional transport operator) resulted in chaos and breakdown of railway transport services in the territory of Silesian Province. In the following weeks, the company faced the possibility of a collapse. The recovery program, implemented by the new local and company authorities at the beginning of 2013, led to gradual improvement both in the field of transport and in the financial situation of the company. At the beginning of 2014, a research team from the University of Economics in Katowice developed a new organizational structure, which was implemented in Koleje Śląskie in mid-2014 [1]. The structure was developed with the application of Henry Mintzberg's structural configuration theory.

Organizational Diagnosis

Koleje Śląskie Sp. z o.o. was established on 17 February 2010 by the self-government of Silesian Province. The company is a so-called internal operator in accordance with the EU Regulation No. 1370 of the European Parliament and of the Council of 23 October 2007 on Public Passenger Transport Services by Rail and by Road. The self-government of the Province is the owner of 100 % of the company. The transport began on 1 October 2011; however, Koleje Śląskie did not take on full responsibilities of an internal operator in the regional market of railway transport until December 2012.

Since the company was not ready from the organizational (including structural) perspective, as well as with regard to the key factors of production (human resources and rolling stock), the consequence was chaos in the transport market, and the company itself reported significant losses. The recovery program undertaken in 2013 consisted of performing the following actions:

- Recapitalization of the COMPANY by the self-government of the Province (in-kind contribution of a railbus with the value of PLN 11.52 million)
- Changes in the area of key resources: human resources (structural changes and increase of the number of engine drivers), rolling stock (termination of lease of a part of the existing rolling stock—15 TRAXX locomotives from Railpool and Lotos Kolej together with engine drivers from 1 June 2013, as well as gradual termination of lease of carriages from Ceske Drahy)
- Reduction of train operation on the railway lines with the lowest profitability
- Organizational changes (establishment of the finance department)
- Implementation of controls of the fundamental economic and operational processes

The analysis of the existing organizational structure of Koleje Śląskie was performed with regard to [2]:

- Specialization of activities (with the purpose of integrating similar activities in organizational units)
- Normalization of activities (concerning procedures used in the organization and expressed in the form of Organizational Regulations specifying the scopes of activity, instructions, and regulations)
- Coordination of activities referring to procedures integrating the functions of departments within the organization (this applies in particular to the procedures related to financial mechanisms, information flow, and risk management)
- Centralization and decentralization of decision making
- Working unit size

The organizational chart presented for reconstruction was developed in August 2013. In particular, the finance and accounting department was added (previously the company had outsourced the activities related to financial services, which is unique in this type of business activity). The organizational structure was a linear structure, the functionalization processes were poorly developed, and official business relations were dominating [3]. Moreover:

- The structure did not contain a separate marketing and commercial department (the units performing such duties reported directly to the Chairman and Vice-Chairman for Transport).
- There was no staff unit (units) handling strategic planning.
- Assignment of certain units was questionable. In particular, this concerned HR Office (within the financial department), ICT Office (within the Chairman's department).

- Nomenclature of units seemed to be random and chaotic (offices, departments, sections, teams).
- Specialization of activities was high; however, the primacy of linearity over functionalization of the organizational structure resulted in poor coordination mechanisms: high centralization of decisions was characteristic; in particular, this concerned the Chairman of the Board and sizes of units were significantly varied (to a large extent, this is justified by the specific character of activities).

Based on the review of the organizational structure, it may be noticed that the particularly important issue which had to be changed was the development of the marketing department, due to the fact that although the company is an internal unit, it is still subject to market pressure. First of all, this is caused by the substitutional impact of individual transport. However, in the territory of Silesian Province, it can be observed that regional railway transport is substituted by regional public transport—especially in the Katowice urban area. Therefore, it is essential to implement the marketing management of Koleje Śląskie, understood as follows [4]:

- Company development in terms of marketing (i.e., recognition of the role of marketing as a key element in the development of the company)
- Regular marketing research
- Marketing planning (with reference to the entire marketing mix, especially products and prices)
- Adjustment of structures, including the emphasis of the role of marketing by establishing the sales and marketing department

It is important to be aware that the process of diagnosing the organization, and in particular the organizational structure, is ongoing. The management and employees of every company are constantly facing the problems of barriers between particular organization units. This is especially visible in linear structures. There are relationships between particular working teams, however, usually with more resistance than in the case of matrix structures, where strong horizontal functional relationships are natural. However, in the case of a railway company, the application of a matrix structure is impossible. Linear structures should be applied particularly in the operational activity, where safety and reliability are the key issues.

Application of Henry Mintzberg's Theory in the Reorganization of the Company

The restructuring process of Koleje Śląskie is well advanced, and therefore it is already possible to undertake activities in the field of developmental restructuring (with planned improvement of the offered services, organization, and finances) [5]. In the first instance, this should concern changes of the organizational structure, which "should follow the strategy," according to A. D. Chandler. Bearing in mind the limited capabilities, as well as the necessity of treating the improvement of the company organization as a process, it is required to indicate model solutions to be followed. One of the most interesting models of ideal organizational structures, based on practical research, is the structural configuration theory, developed by H. Mintzberg in 1979 [6]. This model was applied for preparing the design of the organizational structure of Koleje Śląskie Sp. z o.o.

The theory of H. Mintzberg determines the following components of a broadly understood organizational structure [6]:

- Five main parts of an organization; in the case of Koleje Śląskie, these are:
  - Strategic apex—the management board with its immediate strategic resources: It is necessary to reconstruct the role of the management board, with a clear indication of the decisional hierarchy, distinction between key and current decisions, as well as strengthening of management resources.
  - Operating core—lower level employees in the field of transport: The weaknesses of Koleje Śląskie in this area were the reason of serious operational problems and organizational improvements are necessary.
  - Middle line—managers between the management board and the operating core: Middle lines must be organized according to a uniform standard (division into functional units).
  - Technostructure—employees handling the formulation, update and control of standards, in particular the units handling security, inspections, controlling, and resources: The technostructure of Koleje Śląskie is dispersed and it is necessary to reorganize it and arrange the organizational localization.
  - Support staff—staff, including especially the legal department, market analysis departments, and social staff: It is necessary to carry out redeployment of staff between departments, as well as to rearrange the reporting structure and organizational localization.
- Five coordination mechanisms:
  - Mutual adjustment—consisting in direct, informal agreements between employees from different organizational units: Although less important in rail companies, which are formalized organizations, however, it still occurs and concerns employees of various departments (e.g., commercial or transport department). This mechanism should be observed by the technostructure and middle line in order to transform particularly strong relations of this type into organizational changes and include them in other relevant coordination mechanisms.
  - Direct supervision—managers at different levels provide instructions and control their execution personally. This is a strong dependency in Koleje Śląskie, characteristic for an organization with a linear structure. Reporting changes, and at times standardization, are required.
  - Standardization of work process—its essence lies in preparing the working process by means of instructions, regulations, and guidelines, which is a basic coordination mechanism in rail companies. It is necessary to improve the process in line with the standards determined by Urząd Transportu Kolejowego (Railway Transport Office), as well as by own developed experience of Koleje Śląskie.

  - Standardization of results, which is to leave freedom for contractors—in rail transport very rare.
  - Standardization of skills—requirements for employees are specified, which determines the functioning of the recruitment, evaluation, and staff promotion systems. High organizational positioning of human resources (HR), as well as process standardization, is required in this respect.

- Nine project parameters:

  - Horizontal specialization (expressed with the number and diversity of tasks performed by employees) and vertical specialization (extent to which the content and way of working can be shaped): Narrow horizontal specialization is recommended in the case of lower level positions related to the execution of railway transport process, whereas vertical specialization should be applicable for managerial positions (middle line, strategic apex), as well as technostructure and support staff. Specialization should be reflected in the provisions of the Organizational Regulations.
  - Behavior and process formalization—consisting in providing descriptions of job positions, as well as related tasks and authorizations (including powers of attorney): The level of formalization depends on the job position. In the case of the structures of Koleje Śląskie, the aim should be the compromise of formalization, depending on the job position. It is also necessary to adopt the process approach to formalization with the development of the organization.
  - Training and indoctrination (understood as instilling values and norms based on which the organization wants to develop): As a young organization, Koleje Śląskie should develop a permanent training system model and identify the values and norms to be instilled in the employees.
  - Unit grouping refers to grouping job positions and organizational units according to uniform principles. In Koleje Śląskie, it is necessary to determine the uniform principles of unit grouping, including in particular the determination of territorial organization and grouping nomenclature.
  - Unit size—depends on operating conditions, in particular on specialization: With reference to the operating core of Koleje Śląskie, changes in this field (based on the assessment of team functioning) are likely to be implemented.
  - Action planning and control system—is a response to the question how the results which are to be achieved in particular job positions and teams should be standardized. In Koleje Śląskie, it is necessary to prepare a strategic plan (business plan, operationalized strategy), to be followed by operational planning for particular functions, with the reservation that this process should be developed slowly and without limiting creativity (especially with reference to the strategic apex, middle line, and technostructure).
  - Liaison devices—teams (job positions) facilitating coordination in the matrix system: Although rare in linear structures, they are possible and sometimes even required, e.g., for organizing cooperation with the environment, and implementation of projects.

  - Vertical decentralization—describes the distribution of powers among the management staff: The problem of power delegation concerns in particular the strategic apex, and it is necessary to decentralize certain powers on the basis of a decision, as well as to focus the management board of the company on strategic issues.
  - Horizontal decentralization—determines the distribution of powers in technostructure, operating core, and support units and it consists in delegating to them decision-making competences from the middle line level: Since Koleje Śląskie must develop their technostructure, it will be necessary to introduce competence shifts in the form of horizontal decentralization.

- Four groups of contingency factors:

  - Organization age and size—railway operators are medium and large companies, and mastering railway transport technologies provides an advantage in the form of experience.
  - Technical system—railway transport is a programmed system; therefore formalization and narrow specialization are possible and frequently recommended.
  - Environment—the environment of the internal (passenger) railway operator is stable, with low risk of the occurrence of unexpected dynamic changes. Therefore, there is no need for excessive specialization of coordination mechanisms, and a higher level of decentralization is possible (upon the completion of the recovery program).
  - Power—recognized in terms of ownership, member needs, and popularity of a particular structural solution: It refers to the distribution of power in informal relations, and role of trade unions and experts. The influence of the environment (owners) is significant; in the case of Koleje Śląskie, a strong influence of the owner (self-government of the Province) and the supervisory board appointed by the owner can be observed, which is understandable during the period of organization and execution of the recovery program related to the situation which occurred in December 2012; however, it is ultimately recommended to strengthen the powers of the management board of the company and increase the extent of independence.

The theory of H. Mintzberg also distinguishes between five model structures:

- Simple structure, centralized by the strategic apex, characteristic for young and small organizations
- Machine bureaucracy, with a visible emphasis of technostructure on standardization
- Professional bureaucracy, where the operating core puts a strong emphasis on professionalization
- Divisionalized form, characteristic for a dynamic environment
- Adhocracy, structure dispersed in different markets

Model structures are not equivalent to the actually existing ones and combined solutions are usually found within a particular structure. However, it is worth to establish the starting point and the direction of changes. In the case of Koleje Śląskie, the starting point would inevitably be the simple structure. The direction of evolution related to professionalization required for the implementation of the recovery plan caused the structure to take the form of professional bureaucracy. Professionalization shall be regarded as a counterbalance to the vision of the management, which is perceived as a key factor of competitive advantage [7]; however, in the event of a lack of control, it creates a large risk and may cause far-reaching losses, as evidenced by the experience of Koleje Śląskie. However, the necessity of standardization and the development of technostructure cause that the actual direction of evolution is the mechanical bureaucracy model (although this may evoke negative associations), understood as an organization with a linear structure, formalized, professional, and with a high degree of standardization.

There should be multiple changes in the organization of the company, resulting to a large extent from the symmetrization of processes occurring in the immediate regulatory environment [8]. The main proposed changes in the organizational structure of Koleje Śląskie include the following:

- Clear division of tasks and competences of the management board of the company, and in particular, the establishment of the hierarchy of particular members of the management board.
- Establishment of a separate commercial and marketing department.
- Exposition of technostructure, consisting in separation of support staff for strategic management processes.
- Changes with regard to coordination mechanisms, consisting in redeployment of staff between and within departments.
- Unit grouping, consisting in structuring relations between units at different levels and the related change of nomenclature with reference to the names of organizational units and job positions.
- Redefining outsourcing—the company is currently dominated by consolidation in connection with the recovery program; however, in the long-term perspective, decisions concerning the scope of vertical integration shall be considered in detail.

The abovementioned changes were reflected in the new organizational structure of Koleje Śląskie Sp. z o.o. The organization of the company should be the subject of constant analysis and adjustment works, in accordance with the "Structure follows Strategy" thesis. Henry Mintzberg's structural configuration theory, applied in this paper, might be a useful tool for structure evaluation and reconstruction.

Summary

The project developed by the research team of the University of Economics in Katowice was implemented and Koleje Śląskie will apply the proposed Organizational Chart and Regulations. The application of H. Mintzberg's structural configuration theory has proven to be successful and demonstrated that practical solutions not only can but also should make use of theoretical output, although applying common sense solutions is frequently not obvious under conditions of fast economic changes. The practice of transport management should refer to the theoretical output of economic sciences, including transport management and economy sciences. The authors of the project were surprised to find out how useful and flexible H. Mintzberg's model was. Moreover, it could be applied for regular adaptation of company organization to changes occurring within the company and in its environment. Therefore, the completed work is not a one-off organizational "turnkey" project, but rather a universal tool for organizational improvement.

References

1. *Optymalizacja struktury organizacyjnej Kolei Śląskich.* UE-CBiE, Katowice (2014) / unpublished/.
2. Stoner J.A.S., Wankel Ch.: *Kierowanie*, PWE. Warszawa (1992).
3. Koźmiński *A.K., Piotrowski W. (ed.):Zarządzanie. Teoria i praktyka.* PWN, Warszawa (1997).
4. Koltler Ph.: *Marketing. Analiza, planowanie, wdrażanie i kontrola.* Gebethner & Ska, Warszawa (1994).
5. Stabryła A.: *Zarządzanie strategiczne w teorii i praktyce firmy.* PWN, Warszawa-Kraków (2000).
6. Mintzberg H.: *Structure in 5'S: A synthesis of the research on organization design.* "Management Science" 1980, vol. 26 No. 3.
7. *KraŚnicka T. (ed.):Przywództwo a konkurencyjnoŚć organizacji przedsiębiorczych.* AE, Katowice (2010).
8. Letkiewicz A.: *Samoregulacja w symetryzacji systemów przedsiębiorstw.* Uniwersytet Gdański, Gdańsk (2013).

MIX
Papier aus verantwortungsvollen Quellen
Paper from responsible sources
FSC® C105338

If you have any concerns about our products,
you can contact us on
ProductSafety@springernature.com

In case Publisher is established outside the EU,
the EU authorized representative is:
Springer Nature Customer Service Center GmbH
Europaplatz 3, 69115 Heidelberg, Germany

Printed by Libri Plureos GmbH
in Hamburg, Germany